INTERNATIONAL

PETROLEUM

FISCAL SYSTEMS

AND

PRODUCTION SHARING

CONTRACTS

DANIEL J[

W0009584

PennWell Books

PennWell Publishing Company
Tulsa, Oklahoma

Copyright © 1994

PennWell Publishing Company

1421 South Sheridan Road

P.O. Box 1260

Tulsa, OK 74101

Library of Congress Cataloging-in-Publication Data

Johnston, Daniel

International petroleum fiscal systems and

production sharing contracts/Daniel Johnston

p. cm.

Includes bibliographical references and index.

ISBN 0-87814-426-9

1. Petroleum industry and trade — Taxation.

2. Petroleum industry and trade — Finance.

3. Oil and gas leases.

4. Petroleum — taxation — Law and legislation.

I. Title.

HD9560.8.A1J64 1994

338.2'3282 — dc20   94-33346

                                                                    CIP

CONTENTS

FIGURES

ILLUSTRATIONS

TABLES

ACKNOWLEDGMENTS

This book reflects the influence and effort of a number of people, many of whom may be surprised to know of my sincere appreciation for their inspiration. Some I have not personally met, and others are long gone. To the many who have contributed to the body of knowledge on this delightful subject, I have deep gratitude.

Much of the material and inspiration for this book comes from the courses I teach on this subject. I appreciate those who have attended these courses for their numerous contributions and encouragement.

I would like to thank the following for their reviews, comments, and their helpful suggestions:

James Ahmad	Jim Edwards	Saddique
Richard Barry	Barrie Fowke	Tom Schmidt
Alfred Boulos	Marvin Gearhart	Sue Rhodes Sesso
Horace Brock	Carroll Geddie	Fatmir Shehu
Harry Campbell	Bert Johnston	Malvinder Singh
Joseph Cassinat	Agron Kokobobo	Dennis Smith
Gary Chandler	Hill Mehilli	Selonna Stahle
Ted Coe	Du Quintono	Chuck Thurmond
Ray Darnell	Aziz Radwan	John Vautrain
Stan Dur	Madeliene Reardon	Dale Wetherbee

I'm particularly grateful for the help from Rich Anderson on Chapter 10 and for Michael Darden's review of Chapter 8.

I really appreciate the artwork provided by David Johnston. I have always been proud to be his twin brother.

I sincerely thank Dr. Jimmie Aung Khin for his everlasting inspiration, encouragement, and his help.

I would especially like to express my gratitude to my sweet friend and wife Jill for her work as an editor. I discovered that she had a fabulous talent for this during the work on my first book. Now I would not think of sending anything out without her reviews, comments, and suggestions. Thank you, Jill.

To my wife Jill and our children:
Erik, Lane, Jill Danielle, Julianna, and David,
and to our parents.

INTRODUCTION

The international petroleum industry involves tremendous wealth and power. In many countries petroleum, whether exported or imported, dominates the economy. Natural resources are the crown jewels. Few industries combine such a dramatic contrast between risk and reward. Countries with petroleum resources carefully guard this wealth.

Petroleum taxation is a vital aspect of the industry. Geological, engineering, and financial principles are universal, yet in the realm of taxation, there is added dimension. The subject is so important that understanding at least the basics is mandatory. One of the absolutely first things a geologist, engineer, landman, lawyer, or economist encounters in the international sector is the diversity of fiscal systems. Countries are unique in the way they structure their taxes, and natural resources get special attention. Governments have no control over the gifts of nature, but they do control taxes.

The focus of the book is on the arithmetic and mechanics of the various kinds of fiscal systems—the factors that drive exploration economics. The emphasis is on *practical* aspects of petroleum taxation and industry/government relationships. There is also fertile ground in the philosophy of petroleum taxation. It has changed the industry. Legal and operational aspects of contract/fiscal terms are also examined to provide a foundation in the dynamics of international negotiations.

Both industry and government viewpoints are addressed in this book. A complete grasp of the subject requires an understanding of the dilemmas and concerns of both sides. There are few things more discouraging for a national oil company than an unsuccessful licensing round. Yet prolonged, inconclusive negotiations can be equally frustrating for oil companies.

This book is written for those interested in petroleum taxation and international negotiations. Much of the subject has evolved within just the last 30 years, yet some aspects of taxation are timeless. The terminology has changed over the years and will continue to develop. There is little standardization of terms in the industry, and the abundance of jargon can be rather daunting. The subjects covered in this book are often simple concepts wrapped with buzzwords. A glossary is provided, which should help.

Much of the material provided here was inspired by questions most frequently asked on the subject. The best answers are fortified with specific examples and many are used throughout the book. The summaries and analysis of various fiscal terms and conditions are believed to be accurate, and every effort has been made to gather up-to-date information about the current conditions in the countries cited. Examples of fiscal terms used here are drawn from numerous public sources. Confidential information has been carefully excluded.

Perhaps more effort could be directed toward the cultural aspect of negotiations and doing business in the international arena. Unfortunately it is beyond the scope of this book to cover that ground. It is a fascinating subject. Some of the most gracious and interesting people in the world are found in the international oil business. They inspired this book.

SEMANTICS

Is it a *permit*, a *license*, a *concession*, an *acreage position*, a *contract area*, a *lease*, or a *block*? Sometimes when referring to petroleum operations in a given country, these terms are used interchangeably. However, the term *concession* implies *ownership* or a *freehold interest* of mineral resources. The term has lost ground in the realm of political correctness and the term *royalty/tax system* is commonly substituted.

A government with a production sharing system does not grant concessions. It grants *licenses* or enters into a *contract* with an operator for a given *contract area*. The term *block* is fairly neutral. A company can have a block in the United Kingdom, which has a concessionary system, and another block in Indonesia governed by a PSC.

The semantics of this business get stretched with the common use of the term *fiscal*. Referring to a country's petroleum taxation/contractual arrangement simply as the *fiscal system* is not absolutely correct. It is a matter of convenience that is preferable to saying the petroleum fiscal/contractual system (which would be a step in the right direction, depending upon the country). But it gets clumsy. The term *fiscal* throughout this book is used to encompass all of the legislative, tax, contractual, and fiscal aspects that govern petroleum operations within a sovereign nation/state and its provinces.

The term *mineral* is also used in this book when referring to natural resources. Neither oil nor gas is a mineral, but the term is handy. This book sticks with the prevailing terminology that constitutes the language of the industry today.

Host governments are usually represented by either a national oil company or an oil ministry or both. They are collectively referred to here as the *state* or the *government*. The term *contractor* has specific connotations which are explained later, but for the sake of convenience, this term is used to mean any company operating in the international arena. In this book, *contractor* means operator, contractor, contractor group, or consortium. The term *concessionaire* in reference to an oil company operating in a concessionary system might be appropriate, but it is not part of the industry vocabulary.

2

PETROLEUM FISCAL SYSTEMS

There are more petroleum fiscal systems in the world than there are countries. This is because many countries negotiate terms. Thus one contractor may have different terms than another in the same country. Furthermore, in many countries there are numerous *vintages* of contract in force at any given time as a result of the evolution of the fiscal system.

Some countries use more than one system during transition periods when they are introducing new terms. Some countries offer both concessionary arrangements as well as service or production sharing contracts. Peru has this option. Regardless of the system used, however, the bottom line is a financial issue that boils down to how costs are recovered and profits divided. This leads right to the heart of taxation theory and the concept of economic rent.

ECONOMIC RENT

Economic theory focuses on the *produce* of the earth derived from labor and capital. Rent theory deals with how this produce is divided among the laborers, owners of the capital, and landowners through wages, profit, and rent.

Economic rent in the petroleum industry is the difference between the value of production and the costs to extract it. These costs consist of normal exploration, development, and operating costs as well as an appropriate share of profit for the petroleum industry. *Rent* is the surplus. Economic rent is synonymous with excess profits. Governments attempt to capture as much economic rent as possible through various levies, taxes, royalties, and bonuses.

Figure 2–1 illustrates the allocation of revenues from oil and gas production for costs and the division of profits. Government profit is equal to gross revenue minus costs. This graph shows that governments view the contractor share of profits as a cost. Exploration, development, and operating costs are also viewed this way because the contractor may ultimately recover those costs out of production. What remains is economic rent, if the government has structured an efficient system.

The problem in determining how to capture rent efficiently is that nine out of 10 exploration ventures are unsuccessful. The profit margin for the petroleum industry must be large enough to accommodate the failures. Developing fiscal terms must account for this risk. Present value theory, expected value theory, and taxation theory provide the foundation of fiscal-system design. The contractor entitlement, share of gross revenues, or production will ultimately consist of the recovery of exploration, development, and operating costs as well as some of the profit.

The objective of host governments is to design a fiscal system where exploration and development rights are acquired by those companies who place the highest value on them. In an efficient market, competitive bidding can help achieve this objective. The hallmark of an efficient market is availability of information. Yet exploration is dominated by numerous unknowns and uncertainty. With sufficient competition the industry will help determine

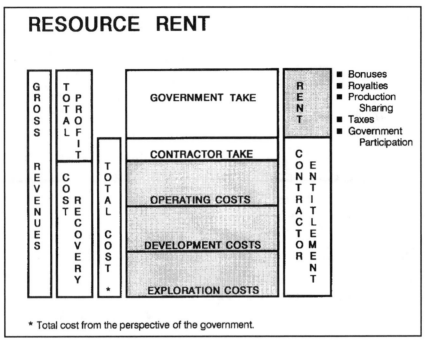

Figure 2-1 Allocations of revenues from production

what the market can bear, and profit will be allocated accordingly. In the absence of competition, efficiency must be designed into the fiscal terms. This is not easy to do.

Governments seek to capture economic rent at the time of the transfer of rights through signature bonuses and during production through royalties, production sharing, or taxes. But production is contingent upon successful exploration efforts. The contractor and government therefore share in the risk that production may not occur. An important aspect as far as risk is concerned is that oil companies are *risk takers*. They can limit risk through diversification. Governments on the other hand are not diversified. They simply are not able to assume as much risk as an international oil company. This is an important dynamic in international negotiations and fiscal design. From the government viewpoint there is a trade-off between risk aversion, where bonuses and royalties are used,

7

and risk sharing, where taxation or production-sharing schemes are used.

A simple bonus bid with no subsequent royalties or taxes would be an extreme example of a government capturing economic rent at the time of transfer of rights. If the government and industry had possession of all the information that ultimately would result from a license before rights were granted, then rent could be clearly defined. The bonus bid would equal the present value of the economic rent. This kind of behavior is seen to some degree in transactions between companies when oil and gas production is purchased and sold. Unfortunately, information about exploration acreage is characterized more by lack of information and uncertainty.

The opposite of a pure bonus bid approach would be pure profit-based taxation. This is more realistic. Governments base most of their taxation on profits, and they are moving even further in this direction.

How governments extract economic rent is important. The industry is particularly sensitive to certain forms of rent extraction, such as bonuses and royalties that are not based on profits. Royalties are of particular concern to the industry because the rate base for royalties is gross revenues.

A spectrum of various elements that make up rent is illustrated in Figure 2–2. The non-profit-related elements of government take, such as royalties and bonuses, are quite regressive—the lower the project profitability, the higher the effective taxes and levies. The further downstream from gross revenues a government levies taxes, the more progressive the system becomes. This is becoming more common. Royalties are being discarded in favor of higher taxes. This has advantages for both governments and the petroleum industry. However, there will always be governments that prefer some royalty. Royalties provide a guarantee that the government will benefit in the early stages of production. There are other ways to do this, and they are discussed further under the concept of *commerciality* in Chapter 4. Additional per-

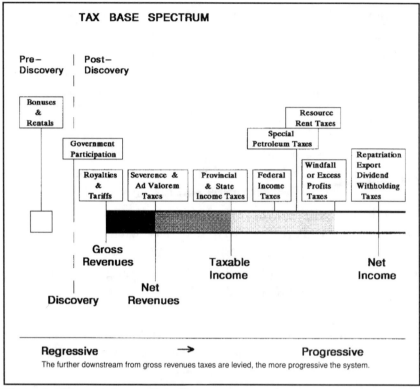

Figure 2–2 The spectrum of taxation

spectives on rent theory are provided in Appendix B.

CONTRACTOR TAKE:
THE COMMON DENOMINATOR

Division of profits boils down to what is called contractor and government *take*. These are expressed as percentages. Contractor take is the percentage of profits to which the contractor is entitled. Government take is the complement of that.

Contractor take provides an important comparison between one fiscal system and another. It focuses exclusively on the division of profits and correlates directly with reserve values, field size thresholds, and other measures of relative economics.

CONTRACTOR TAKE: SYNONYMS	GOVERNMENT TAKE: SYNONYMS
• Company take	• State take
• Contractor marginal take	• Government marginal take
• Contractor share of profits	• Government share of profits
• Contractor after-tax equity split	• Government after-tax equity split

Under a system such as the Indonesian production-sharing contract with the well-known 85%/15% split in favor of the government, the contractor may still end up with a 35%–50% share of production. This is because of the reimbursement of costs, or what is known as *cost recovery*, which is discussed in detail in Chapter 4. The Indonesian 15% contractor take is a measure of the contractor's share of profits. It is a meaningful number. The formulas for calculating take are as follows:

Operating income ($) = Cumulative gross revenues minus cumulative gross costs over life of the project

Government income ($) = All government receipts from royalties, taxes, bonuses, production, or profit sharing, etc.

Government take (%) = Government income ÷ operating income

Contractor take (%) = 1 – Government take

The best way to calculate take requires detailed economic modeling using cash flow analysis. Once a cash flow projection has been performed, the respective takes over the life of the project can be evaluated. An example is shown in Table 2–1. It is assumed that these values have been retrieved from a cash flow projection. The gross revenues are $1,000, and costs over the life of the project equal $400. The total profit, therefore, is $600. The contractor share of profits is $250. This is equal to 42% of the

Table 2-1

Calculation of Government/Contractor Take

Gross revenues	$1,000	
Total costs	–400	Capital and operating
Operating cncome	600	Total profits
Royalties and taxes	$350	Government share
Net after-tax income	$250	Contractor share
Contractor take	42%	($250/$600)
Government take	58%	($350/$600)

profits. Contractor take is 42%. Government take is 58%.

Under systems with liberal cost recovery provisions, the government take comes at a relatively later stage of production than under those systems with royalties and restrictions on cost recovery. An even more detailed analysis of the respective takes would include the present value dimension. But this is not ordinarily done.

A method is outlined here for estimating government/contractor take without detailed cash flow modeling. There are limitations to a quick-look approach, but 95% of the time, estimates of contractor/government take provide extremely valuable information. It is not difficult to estimate.

The main limitation of estimating take is that it is not always easy to account for other aspects of a given fiscal system, such as cost recovery limits, investment credits, royalty or tax holidays, and domestic market obligations (DMOs). These are explained later in Chapter 4.

Estimating Contractor Take Step-by-Step

1. It is convenient to work with percentages. Start with 100%, which represents all gross revenues, and subtract the

royalty percentage. This gives percentage net revenue.

2. Estimate the overall development and operating costs; for instance, 35%. Deduct this percentage from net revenue.

Over the life of a project, these costs range from 20%–40% of gross revenues. If costs exceed this level, then it is likely a project will be submarginal, depending on the fiscal system. A quick-look estimate should focus on a hypothetically profitable venture. Companies normally do not develop subeconomic fields. A cost estimate of 35% of gross revenues is a good place to start and provides a common point of reference.

3. Subtract taxes, levies, government profit share, etc. This gives the contractor's share of profits.

4. Subtract the percentage of costs from 100% (gross revenues). This equals total profits.

5. Divide the contractor share of profits after income tax by total profits. This is the contractor take.

For example, assume a system with a simple 15% royalty and 40% income tax. Costs equal 35% of gross revenues. In this example, the contractor ends up with 46.15% of profits—contractor take equals 46.15%.

Step 1	100%	Gross revenues
	– 15	Royalty (15%)
	85%	Net revenue percentage
Step 2	– 35	Costs
	50%	Taxable revenues
Step 3	– 20	Income tax (40%)
	30%	Contractor after-tax share
Step 4	100%	(gross revenues) – 35% costs = 65% profits
Step 5	30% ÷ 65% = 46.15%	
	= Contractor take	

The fiscal terms of a number of countries are included in Appendix A. The contractor take, cost recovery, and government participation of selected countries is summarized in Table 2–2. These contractor take figures assume a level of costs on the order of 35%.

Implications of Contractor Take

In 1993 an independent U.S. oil company made a 25-MMBBL discovery off the Northwest Shelf of Western Australia. The discovery was not exactly headline news outside Western Australia. The fiscal terms there are quite good though: over three times better than in Indonesia or Malaysia. The discovery therefore amounted to the equivalent of an 80-MMBBL Indonesian or Malaysian field. The fiscal terms make a *huge* difference.

The level of contractor take also has a direct impact on the value of reserves. Based on a wellhead price of $18.00/bbl, the value of proved, developed, producing (PDP) working-interest reserves in Indonesia is $1.10–$1.60/bbl. In the United States, the value of similar reserves would be $4.50–$6.00/bbl. This relationship is illustrated further in Figure 2–3.

Valuation Rule of Thumb

Proved, developed, producing reserves are worth from one-half to two-thirds of the wellhead price times the contractor's take.

Like any rule of thumb, this one should be used with caution. It is used to make a quick estimate of the present value of a contractor's working-interest share of proved, developed, producing reserves, assuming there are no major sunk costs available for cost recovery. Proved, undeveloped (PUD) reserves will be worth much less. The value of undeveloped reserves is usually less than half of the value of PDP reserves.

An important distinction is made here in reference to the value of reserves in the ground. The previous example and the one in Figure 2–3 are based upon the contractor's working-interest share of reserves, not the contractor's *entitlement*. This subject is developed further later in the book, particularly in Chapter 4. Unless

Table 2–2

Comparison of Terms—Selected Countries

Country	Contractor Take, % *	Cost Recovery Limit, %	Maximum Government Participation, %
Abu Dhabi (OPEC Terms)	9–12	100	0
Albania	20–25	45	0
Angola	20	50	50
Australia	40–50	100	0
Brunei	28–30	100	50
Cameroon	14–16	?	50
China	38–41	50–60	51
Colombia	30–37	100	51
Congo	30–35	100	50
Egypt	24–28	30–40	50
Gabon	20–25	40–55	10
India	30–42	100	30
Indonesia	11–13	80	15[1]
Indonesia E.	30–33	80	15[1]
Ireland	75	100	0
Korea	36–40	100	0
Malaysia	14–19	50–60	15
Morocco	40–44	100	0
Myanmar	21–23[2]	40	0
New Zealand	47–51	100	0
Nigeria	10–18	40	?
Norway	18	100	?
Papua New Guinea	30–35	100	22.5
Philippines	44–47	70	0
Spain	60	100	0
Syria	18–22	25–35	0
Thailand	30–44	100	?
Timor Gap	26	90	0
United States	42–53	AMT[3]	0
Vietnam	30	40	0

[1]Indonesia seldom exercises its right to participate.
[2]Tax holiday on first three years' production
[3]Alternative Minimum Tax
* Excludes Government participation — usually a carried interest.

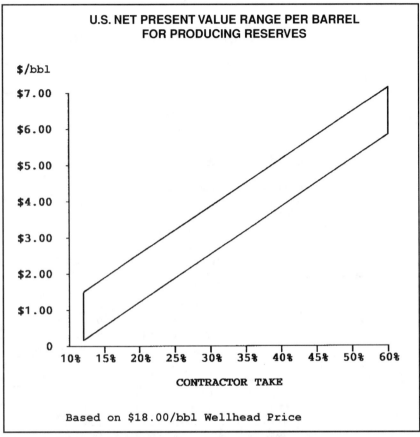

Figure 2–3 Comparing fiscal terms—FMV per barrel

otherwise stated, all references to reserve volumes or field sizes deal either with gross recoverable reserves or reserves attributable to a company's working interest.

The economic perspective of the exploration, development, and production phases can be quite different. Often the exact context must be indicated. For example, when discussing threshold field size, there is a huge difference between exploration and development thresholds. Most of the science of fiscal-system analysis deals with exploration economics and, to a lesser extent, development and production economics. Therefore the changing perspectives are a key consideration. Table 2–3 illustrates aspects

Table 2–3

Changing Economic Perspectives

	Exploration Economics	Development Economics	Production Economics
Focus	Risk Analysis Farm-in/Farm-out Bidding	Feasibility Exploitation Evaluation	Evaluation Acquisitions
Risk	Drilling/Finding **Very High**	Costs, timing, prices **Moderate**	Mostly price risk **Lower**
Bonuses	Signature Bonus **Very Sensitive**	Startup Bonus Usually not sensitive	Production Bonuses Usually not too sensitive
Work Commitment	Very important part of risk capital	Huge but not as risky as exploration	Usually not applicable
Key Capital Requirements	Exploration Capital Seismic, Bonuses G&G, Drilling	Development Costs Drilling, Facilities, Transportation	Operating Costs
Operating Costs	Economics not sensitive	Moderately sensitive	Very sensitive
Sunk Costs	Not yet created	Positive impact on development decision	Flow through cost recovery and have present value
Threshold Field Size	Huge 100 – 500 + MMBBLS	Moderate 5 – 50 + MMBBLS	Economic Limit Threshold
Contractor Take	Critical concern	Important concern	Important concern
Royalties	Can be very regressive and discourage exploration and development, and cause premature abandonment		
Cost Recovery Limit	Important concern	Very important concern	Not so important
Reserve Values	Hypothetical Expected Value	Low Discounted Cash Flow	Highest Discounted Cash Flow

that are subject to change and how they are viewed. Everything is relative.

NEGOTIATIONS

Governments have devised numerous frameworks for extracting economic rent from the petroleum sector. Some are well balanced, efficient, and cleverly designed. Some will not work. The fundamental issue is whether or not exploration and/or development is feasible under the conditions outlined in the fiscal system. The result of government efforts are sometimes referred to as *fiscal marksmanship*—either poor or good. Structuring a fiscal system that will be appropriate or *on target* under a variety of unknown future circumstances is nearly impossible.

The purpose of fiscal structuring and taxation is to capture *all* economic rent. This is consistent with giving the industry a reasonable share of profit, or take. But the level of industry profit considered to be fair and reasonable is debatable. The issue of the division of profits lies at the heart of contract/license negotiations.

GOVERNMENT OBJECTIVES

The objective of a host government is to maximize wealth from its natural resources by encouraging appropriate levels of exploration and development activity. In order to accomplish this, governments must design fiscal systems that

- Provide a fair return to the state and to the industry
- Avoid undue speculation
- Limit undue administrative burden
- Provide flexibility
- Create healthy competition and market efficiency

The design of an efficient fiscal system must take into consideration the political and geological risks as well as the potential rewards.

Malaysia has one of the toughest fiscal systems in southeast Asia. But Malaysia has good geological potential. Many companies would love to explore in Malaysia, and the government knows this. Governments are not the only ones who draw the line between fair return and rent. The market works both ways.

One country may tax profits at a rate of 85% or more, like Indonesia, while another country may have an effective tax rate of only 40%, like Spain. Yet both countries may be efficiently extracting their resource rent regardless of the kind of system that is used.

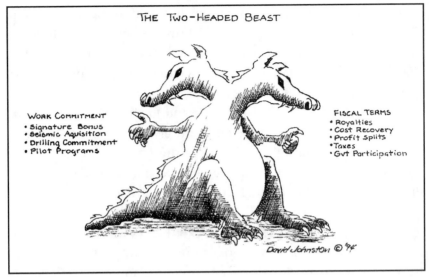

THE TWO-HEADED BEAST

WORK COMMITMENT
• Signature Bonus
• Seismic Aquisition
• Drilling Commitment
• Pilot Programs

FISCAL TERMS
• Royalties
• Cost Recovery
• Profit Splits
• Taxes
• Gvt Participation

David Johnston © '94

Illustration 2–1

OIL COMPANY OBJECTIVES

The objectives of oil companies are to build equity and maximize wealth by finding and producing oil and gas reserves at the lowest possible cost and highest possible profit margin. In order to do this, they must search for huge fields. Unfortunately, the regions where huge fields are likely to be found are often accompanied by tight fiscal terms. The oil industry is comfortable with tough terms if they are justified by sufficient geological potential. This is the birthplace of dynamic negotiations.

The primary economic aspects of contract/license negotiations are the work commitment and the fiscal terms. These are sometimes collectively referred to as the *commercial* terms. The challenge that faces negotiators is the two-headed beast shown in Illustration 2–1. The work commitment represents hard *risk dollars* while fiscal terms govern the allocation of revenues that may result from successful exploration efforts. If negotiators can get past this beast, the legal terms are right behind it.

Figure 2–4 shows how these elements influence the basic industry risk/reward relationship. This is a graphical representation

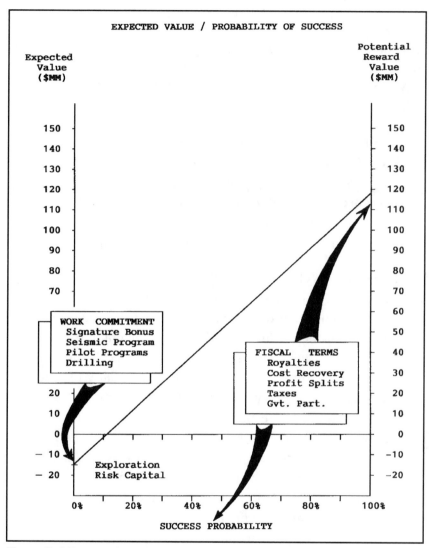

Figure 2–4 Expected monetary value graph—risk/reward

of a simple two-outcome Expected Monetary Value Theory (EMV) risk model. The work commitment dominates the *risk* side of the equation, while the fiscal terms influence the success ratio estimate as well as the *reward* side of the equation. The equation represented by the graph in Figure 2–4 is shown here:

EXPECTED MONETARY VALUE (EMV)

EMV = (Reward $\times$ SP) – [Risk capital $\times$ (1 – SP)]
where

EMV = Expected monetary value
Risk capital = Bonuses, dry hole costs, G&G, etc.
SP = Success probability
Reward = Present value of a discovery based on
discounted cash flow analysis
discounted at corporate cost of capital.

The equation effectively yields a weighted average. It is composed of the value of a possible discovery multiplied by the chance of making that discovery minus the risk capital times the probability that there will be no discovery. The graph or the equation can be used in different ways. For example, if the dry-hole cost for a prospect is $15 million and the probability of success is 12.5%, then the potential reward better be worth at least $120 million or more, or the prospect is not worth consideration. Under some of the tougher fiscal systems in the world where the present value of proved, undeveloped reserves is less than 50¢/bbl, the required recoverable reserves for a prospect like this would exceed 240 MMBBLS. But if the risk capital were cut in half, the required target would also be cut in half.

If the discounted present value of a prospect is $120 million and the dry-hole costs are $15 million, then the prospect had better have a probability of success of greater than 12.5%. This is the break-even success ratio. If the company believed this prospect had a 20% chance of success, then the expected value would be $12 million.

EMV = **(Reward $\times$ SP) – [Risk capital $\times$ (1 – SP)]**
= ($120 MM $\times$ .20) – [$15 MM $\times$ (1 – .20)]
= $12 MM

The expected value is influenced by both the risk capital and the fiscal terms that partially govern the size of the reward. Notice that the risk capital weighed much more heavily in the decision-making process. Because of the 20% chance of success, the risk dollars outweighed the reward dollars 4:1. With a success ratio of around 10%, the relationship changes to 9:1. Risk dollars are critical. This is why so much attention is placed on work commitments and bonuses. Bonuses in particular are hard for companies to bear because they provide no information. At least a seismic program or a drilling program adds to the body of information about an area.

FAMILIES OF SYSTEMS

Governments and companies negotiate their interests in one of two basic systems: concessionary and contractual. The fundamental difference between them stems from different attitudes towards the *ownership* of mineral resources. The Anglo-Saxon and the French concepts of ownership of mineral wealth are the root beginnings. This ownership issue drives not only the language and jargon of fiscal systems, but the arithmetic as well. The classification of petroleum fiscal systems is outlined in Figure 2–5.

CONCESSIONARY SYSTEMS

Concessionary systems, as the term implies, allow private ownership of mineral resources. The United States, of course, is the extreme example of such a system where individuals may own mineral rights. This concept of ownership comes from Anglo-Saxon legal tradition. In most countries the government owns all mineral resources, but under concessionary systems it will transfer title of the minerals to a company if they are produced. The company is then subject to payment of royalties and taxes.

CONTRACTUAL SYSTEMS

Under contractual systems the government retains ownership of minerals. Oil companies have the right to receive a share of

21

production or revenues from the sale of oil and gas in accordance with a production sharing contract (PSC) or a *service* contract.

Production sharing is rooted in the Napoleonic era French legal concept of the ownership of minerals—that mineral wealth should not be owned by individuals but by the state for the benefit of all citizens. Indeed, this philosophy is embodied in the 1945 Indonesian Constitution Article 33, which states

> All the natural wealth on land and in waters is under the jurisdiction of the State and should be used for the benefit and welfare of the people.

In the petroleum industry, Indonesia is the pioneer of the PSC, with the first contracts signed in the early and mid-1960s. Indonesia is the standard of comparison for all PSCs. Accordingly, the Indonesian PSC is given special attention in this book. In France, where the notion of state ownership of mineral wealth ultimately inspired PSCs, the petroleum fiscal system is not PSC based. France has a royalty/tax system—a bit of irony.

It is partially because of the concept of ownership of mineral resources that the term *contractor* has come into such wide use. The earliest uses of the production sharing concept occurred in the agriculture industry. Therefore the term is used in the same context as *sharecropper*, where ownership of the land and minerals is held by the government/landlord. The contractor or tenant/sharecropper is compensated out of production of minerals or grain according to a specific sharing arrangement. The term *contractor* therefore applies to PSCs or service agreements only, but with practical usage it cuts across the boundaries between PSCs and concessionary systems.

There is another aspect to this ownership issue. In most contractual systems, the facilities emplaced by the contractor within the host government domain become the property of the state either the moment they are landed in the country or upon startup or commissioning. Sometimes title to the assets or facilities does not pass to the government until the attendant costs have been recovered. This transfer of title on assets, facilities, and equipment does not apply to leased equipment or to equipment brought in by service companies.

Contractual arrangements are divided into *service contracts* and *production sharing contracts*. The difference between them depends on whether or not the contractor receives compensation in *cash* or in *kind* (crude). This is a rather modest distinction and, as a result, systems on both branches are commonly referred to as PSCs, or sometimes production sharing agreements (PSAs).

For example, in the Philippines the government alternately refers to their contractual arrangement as either a service contract or a PSC. The oil community does the same thing, but more ordinarily refers to this system as a PSC. In a strict sense though, this system is a *risk service contract* because the contractor is paid a fee for conducting exploration and production operations.

From a legal point of view, the timing of the transfer of title and ownership is important. If disputes arise, the closer the contractor is to actual physical ownership, the stronger the legal position. As far as ownership is concerned, under a PSC the contractor does ultimately receive a share of production and hence takes title to the crude. The transfer of title is effectively shifted from the wellhead under a concessionary system to the point of export under a PSC. With a service or risk service agreements, the issue of ownership is removed altogether. In a service contract, the contractor gets a share of *profits*, not production. With that in mind, the term *revenue sharing* or *profit sharing* contract would be appropriate, but these terms are not used. However, under some service agreements, the contractor has a right to purchase crude from the government at a discount, so ultimately the contractor ends up with title to some crude. These ambiguities in ownership and timing of ownership cloud the issue of categorization. In fact, there are numerous systems that practically defy classification, and the fundamental issue of *ownership* is the last resort. However, categorization is not the important thing. Economics is.

Despite the differences between systems it is possible to obtain the same economic results under a variety of systems. There are so few material differences between one family and another as far as the calculations are concerned that it is difficult to make generalizations about any given family of systems.

The difference between risk service and pure service contracts depends on whether the fee is based on profits or not. Pure service contracts are quite rare. In pure (nonrisk) service contracts the contractor carries out exploration and/or development work on behalf of the host country for a fee. All risk is borne by the state. This arrangement is characteristic of the Middle East where the state often has substantial capital but seeks outside expertise and/or technology.

Pure service agreements, rare as they are, can be quite similar to arrangements found in the oil service industry. The contractor is paid a fee for performing a service—no risk element. In the late 1950s, the Argentine government under President Arturo Frondizi negotiated a number of service contracts known as "The Frondizi Contracts." These contracts were negotiated with oil companies for drilling, development, and medium-risk exploration services. The companies included Kerr McGee, Marathon, Shell, Esso, Tennessee Gas Transmission, Cities Service, Amoco, and Union Oil. The drilling service contracts were pure service arrangements whereby the contractor was paid on a footage basis for drilling services and an hourly basis for testing and completion operations. The payment was usually a combination of dollars and pesos.

While many service agreements are identical to PSCs in all but the method of payment, many have unique contract elements that are used in calculating the service fee. Some of these can get rather exotic and may indicate new directions in fiscal design. These are discussed in Chapter 5.

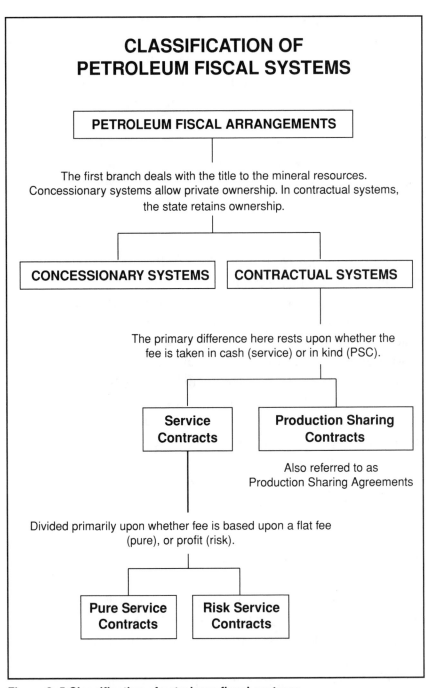

Figure 2–5 Classification of petroleum fiscal systems

VARIATION ON TWO THEMES

Numerous variations and twists are found under both the concessionary and contractual themes. The philosophical differences between the two systems have fostered a terminology unique to each. However, the terms are simply different names for basic concepts common to both systems.

Because of the modest difference between service agreements and PSCs, the study of PSCs effectively covers the whole contractual branch of the tree. Much of the language and arithmetic of PSCs and service agreements is identical. PSCs outnumber service agreements 5:1. As of 1994 there were only nine countries using service agreements compared to 44 using PSCs.

Comparing and contrasting PSCs with concessionary systems provides solid foundation for the study of petroleum fiscal systems. This book concentrates on the terminology and arithmetic of these two families. Most of the differences—especially from a practical and financial point of view—are purely semantic. The most dramatic differences between one fiscal system and another have to do with just how much taxation is imposed.

In addition to the two main families, there are a few arrangements that appear to be a type of fiscal system. Each of them are discussed in detail in Chapter 5. They include
- Joint ventures
- Technical assistance contracts, EOR contracts
- Rate of return contracts

Joint ventures (JVs) are not a type of fiscal/contractual system. They are common in the industry through standard joint operating agreements (JOAs) and working-interest arrangements between companies. Governments also get directly involved through joint ventures. The term is primarily used to describe arrangements where the national oil company is in partnership with the contractor. Government/industry joint ventures are also referred to as government participation. This is found in both concessionary and contractual systems. However, some joint venture arrangements so strongly characterize the operating environment that special attention is given to them in Chapter 5.

Technical assistance contracts (TACs) are used for enhanced oil recovery (EOR) projects or rehabilitation/redevelopment schemes administered under a PSC or a concessionary system. These normally involve proved reserves that are beyond the primary recovery stage. This creates quite a different setting than normal exploration and development economics. The critical aspect of exploration risk is missing.

Rate of return (ROR) features are also found in both systems. ROR is more a descriptive term to identify further the nature of a particular system. For example, the Papua New Guinea (PNG) fiscal system is a concessionary-based ROR system. Equatorial Guinea, on the other hand, uses a PSC-based ROR contract. The ROR concept can get slightly exotic. And it appears to be a new direction for the industry.

Specific examples of JVs, TACs, and ROR contracts are given in Chapter 5.

CONCESSIONARY SYSTEMS

Concessionary arrangements predominated through the early 1960s. The earliest agreements consisted of only a royalty payment to the state. The simple royalty arrangements were followed by larger royalties. Taxes were added once governments gained more bargaining power. In the late 1970s and early 1980s, a number of governments created additional taxes to capture excess profits from unexpectedly high oil prices. Now there are numerous fiscal devices, layers of taxation, and sophisticated formulas found in concessionary systems.

FLOW DIAGRAM

Figure 3–1 depicts revenue distribution under a simple concessionary system. This example is provided to develop further the concept of contractor take and to compare with other systems. The diagram illustrates the hierarchy of royalties, deductions, and (in this example) two layers of taxation. For illustration purposes, a single barrel of oil is forced through the system.

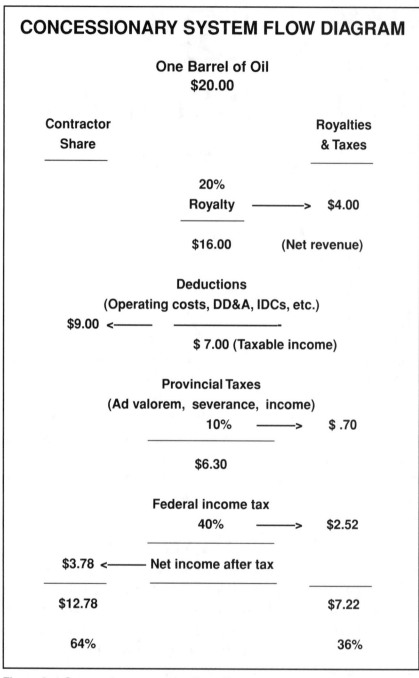

Figure 3–1 Concessionary system flow diagram

FIRST: ROYALTIES

The royalty comes right off the top. In this example a 20% royalty is used. Gross revenues less royalty equals net revenue.

SECOND: DEDUCTIONS

Operating costs; depreciation, depletion, and amortization (DD&A); and intangible drilling costs (IDCs) are deducted from net revenue to arrive at taxable income. DD&A is the common terminology, but depletion is seldom allowed. When the term is used, it is assumed that it applies to depreciation and amortization. Most countries follow this format but will have different allowed rates of depreciation or amortization for various costs. Some countries are liberal in allowing capital costs to be expensed and not forced through DD&A.

THIRD: TAXATION

Revenue remaining after royalty and deductions is called taxable income. In this example, it is subjected to two layers of taxation: 10% provincial and 40% federal taxes. Provincial taxes are deductible against federal taxes. The effective tax rate therefore is 46%.

With tax deductions the contractor share of gross revenues is 64%. The profit in this example is $11.00 ($20.00 minus $9.00 in costs). The contractor's share of profits is $3.78. Contractor take therefore is 34.36%. This is different than profit margin, which in this example is 18.9% ($3.78/$20.00).

CASH FLOW PROJECTION

Table 3–1 outlines further the terminology and hierarchy of arithmetic for calculating contractor cash flow. This example is more of a financial perspective, but the arithmetic is consistent with the basic equations that are included in this chapter. Table 3–2 is a cash flow projection for a concessionary system. It includes a detailed explanation of the operations involved in arriving at a year-by-year cash flow calculation. The royalty in this example is 12.5%, and the income tax rate is 35%. This is the usual format for cash flow analysis of a possible field development. In this case, the discovery is assumed to be a 40-MMBBL field, and production is profiled each year. The last column (N) is the contractor net after-tax cash flow. The next steps in economic modeling are to calculate the present value of the cash flow stream and to estimate the internal rate of return (IRR).

The cash flow projection is based on the assumption that some classes of capital cost are intangible and are immediately deductible. The tangible capital costs are depreciated over five years.

The development scenario outlined in this projection is for a 40-MMBBL field with a projected 11-year life. Total capital costs are $101 million, and estimated operating costs during the life of the project are $117 million. The following calculation of the respective takes comes from the cash flow projection. The contractor take is 53.3%. A quick-look estimate using a 35% cost assumption yields a contractor take of 52.5%.

CALCULATION OF GOVERNMENT/CONTRACTOR TAKE

Gross revenues	$719,877	
Total costs	– 217,993	
Total profit	$501,884	
Royalties	– 89,985	
Taxes	– 144,165	
Contractor take	$267,734	
Contractor take	53%	($267,734 ÷ $501,884)
Government take	47%	

BASIC EQUATIONS, ROYALTY/TAX SYSTEMS

Gross revenues   =   Total oil and gas revenues

Net revenues   =   Gross revenues
                               −   royalties

Net revenue (%)   =   100% − Royalty rate (%)

Taxable income   =   Gross revenues
                               −   Royalties
                               −   Operating costs
                               −   Intangible capital costs[1]
                               −   DD&A (including abandonment costs)
Deductions   −   Investment credits (if allowed)
                               −   Interest on financing (if allowed)
                               −   Tax loss carry forward
                               −   Bonuses[2]

Net cash flow   =   Gross revenues
(aftertax)   −   Royalties
                               −   Tangible capital costs
                               −   Intangible capital costs[1]
                               −   Operating costs
                               −   Bonuses
                               −   Taxes

[1] In many systems no distinction is made between operating costs and intangible capital costs, and both are expensed.
[2] Bonuses are not always deductible.

Table 3-1

Concessionary System Structure		
Oil Company Perspective		
Terminology	$/bbl	Royalties, Costs, and Taxes
Wellhead price	$20.00	
	−3.75	18.75% ($^3/_{16}$) Royalty
Net revenue	16.25	
	−1.63	10% Severance, ad valorem, and production taxes
	−4.15	Operating costs
	−1.45	General & administrative costs
Before-tax operating income	9.02	
	−5.15	Depreciation, depletion, & amortization
Before-tax net income	3.87	
	− .31	8% State income tax
	3.56	
	−1.21	34% Federal income tax
After-tax net income	$2.35	
	+5.15	Depreciation, depletion, & amortization
	−1.25	Tangible capital costs
After-tax cash flow	$6.25	

Table 3-2

Sample Royalty/Tax System Cash Flow Projection

Year	Oil Production (MBBLS) (A)	Oil Price ($/bbl) (B)	Gross Revenues ($M) (C)	12.5% Royalty ($M) (D)	Net Revenues ($M) (E)	Intangible Cap Ex ($M) (F)	Tangible Cap Ex ($M) (G)	Operating Expense ($M) (H)	DD&A ($M) (I)	Total Applied Deductions ($M) (J)	Tax Loss C/F ($M) (K)	Taxable Income ($M) (L)	35.0% Income Tax ($M) (M)	Net Cash Flow ($M) (N)
1994	0	$18.00	0	0	0	10,000	10,000	0	0	0	0	(10,000)	0	(20,000)
1995	0	18.00	0	0	0	5,000	8,000	0	0	0	10,000	(15,000)	0	(13,000)
1996	0	18.00	0	0	0	3,000	40,000	0	0	0	15,000	(18,000)	0	(43,000)
1997	4,500	18.00	81,000	10,125	70,875	0	25,000	11,500	16,600	46,100	18,000	24,775	8,671	25,704
1998	7,000	18.00	126,000	15,750	110,250	0	0	14,000	16,600	30,600	0	79,650	27,878	68,373
1999	5,600	18.00	100,800	12,600	88,200	0	0	12,600	16,600	29,200	0	59,000	20,650	54,950
2000	4,760	18.00	85,680	10,710	74,970	0	0	11,760	16,600	28,360	0	46,610	16,314	46,897
2001	4,046	18.00	72,828	9,104	63,725	0	0	11,046	16,600	27,646	0	36,079	12,627	40,051
2002	3,439	18.00	61,904	7,738	54,166	0	0	10,439	0	10,439	0	43,727	15,304	28,422
2003	2,923	18.00	52,618	6,577	46,041	0	0	9,923	0	9,923	0	36,118	12,641	23,477
2004	2,485	18.00	44,725	5,591	39,135	0	0	9,485	0	9,485	0	29,650	10,378	19,273
2005	2,087	18.00	37,569	4,696	32,873	0	0	9,087	0	9,087	0	23,786	8,325	15,461
2006	1,732	18.00	31,183	3,898	27,285	0	0	8,732	0	8,732	0	18,552	6,493	12,059
2007	1,427	18.00	25,570	3,196	22,374	0	0	8,421	0	8,421	0	13,953	4,884	9,069
2008	0	18.00	0	0	0	0	0	0	0	0	0	0	0	0
	40,000		719,877	89,985	629,893	18,000	83,000	116,993	83,000	217,993		368,899	144,165	267,735

(A) Production Profile (Thousands of barrels per year)
(B) Crude Price ($/bbl)
(C) Gross Revenues = (A) x (B)
(D) Royalty = 12.5% = (C) x .125
(E) Net Revenues = (C) – (D)
(F) Intangible Capital Costs [Expensed]
(G) Tangible Capital Costs [Capitalized, see Column (I)]
(H) Operating Costs [Expensed]
(I) Depreciation of Tangible Capital Costs: 5-Year Straight Line Decline
(J) Total Applied Deductions = If (F) + (H) + (I) + (K) is greater than or = to (E) then (E) Otherwise (F) + (H) + (I) + (K)
(K) Tax Loss Carry Forward = If (L) from previous year is negative then it is brought forward otherwise (zero)
(L) Taxable Income = (E) – (F) – (H) – (I) – (K)
(M) Income Tax = 35% = If (L) is positive .35 x (L), otherwise (zero)
(N) After Tax Net Cash Flow = (C) – (D) – (F) – (G) – (H) – (M)

Another dimension of contractor take comes from the effect of royalties or any taxes that are levied on gross revenues instead of profits. With different levels of profitability, fiscal systems with royalties can yield different government/contractor takes.

The contractor take calculation using the U.S. federal offshore example illustrates the impact of a royalty. The step-by-step allocation of revenues under a high- and low-cost case is based on the following assumptions:

Gross revenues   = $100 million
Royalty = $1/6$ =   16.67%
Costs eligible for deduction =
$50 million – high cost case
$20 million – low cost case

U.S. FEDERAL OFFSHORE FISCAL STRUCTURE

Low Cost Case	High Cost Case	
$100.00 MM	$100.00 MM	Gross revenues
–16.67	– 16.67	Royalty
83.33	83.33	Net revenues
–20.00	–50.00	Deductions
63.33	33.33	Taxable income
–21.53	–11.33	**Federal income tax 34%**
41.80	22.00	Contractor net income after tax
+20.00	+50.00	Deductions
$61.80 MM	$72.00 MM	**Contractor total**
52.25%	**44.0%**	**Contractor take**
		Contractor net income after tax ÷ Gross revenues – Costs

The government share of revenues in the low-cost case is only $38.2 million, or 38.2%. The contractor's share of gross revenues was $61.8 million. The taxable profit is $80 million ($100 MM – $20 MM deductions). Of this, the contractor's share is $41.8 million. The contractor take therefore is 52.25% ($41.8 MM/$80 MM). Contractor take under the high-cost regime is only 44.0%.

This is a regressive fiscal structure. The lower the profitability, the higher the effective tax rate. This is because of the royalty. It is based on gross revenues.

PRODUCTION SHARING CONTRACTS

At first PSCs and concessionary systems appear to be quite different. They have major symbolic and philosophical differences, but these serve more of a political function than anything else. The terminology is certainly distinct, but these systems are really not that different from a financial point of view. There is a general view that PSCs are more complex and onerous than concessionary systems. This is not a fair generalization. Too much diversity exists on both sides.

The arithmetic of a simple PSC is evaluated first. Many of the other features of a PSC are similar to those found under other systems. Therefore, these common elements are discussed in detail later.

In the numerous production sharing arrangements, there are common elements. The essential characteristic, of course, is that of state ownership of the resources. The contractor receives a share of production for services performed.

As more countries open their doors to the petroleum industry, they are using the PSC as opposed to concessionary systems. The first PSC was signed by IIAPCO in August 1966 with Permina, the Indonesian National Oil Company at that time (now Pertamina). This is when oil companies started becoming contractors. This contract embodied the basic features of the production sharing concept:

- Title to the hydrocarbons remained with the state.
- Permina maintained management control, and the contractor was responsible to Permina for execution of petroleum operations in accordance with the terms of the contract.
- The contractor was required to submit annual work programs and budgets for scrutiny and approval by Permina.
- The contract was based on production sharing and not a profit-sharing basis.
- The contractor provided all financing and technology required for the operations and bore the risks.
- During the term of the contract, after allowance for up to a maximum of 40% of annual oil production for recovery of costs, the remaining production was shared 65%/35% in favor of Permina. The contractor's taxes were paid out of Permina's share of profit oil.
- All equipment purchased and imported into Indonesia by the contractor became the property of Permina. Service company equipment and leased equipment were exempt.

These features continue to outline the nature of the government/contractor relationships under PSCs or service agreements. It is a formula that is popular with many governments.

FLOW DIAGRAM

Figure 4–1 shows a flow diagram of a PSC with a royalty. It illustrates the terminology and arithmetic hierarchy of a typical PSC. For illustration one barrel of oil is used.

FIRST: ROYALTY

The royalty comes right off the top just as it would in a concessionary system. This example uses a 10% royalty.

BASIC EQUATIONS, CONTRACTUAL SYSTEMS

Gross revenues = Total oil and gas revenues

Net revenues = Gross revenues – Royalties

Net revenue (%) = 100% – Royalty rate (%)

Cost recovery = Operating costs
"Cost oil"   + Intangible capital costs[1]
+ DD&A (including abandonment costs)
+ Investment credits (if allowed)
+ Interest on financing (if allowed)
+ Unrecovered costs carried forward

Profit oil = Net revenue – Cost recovery

Contractor profit oil = Profit oil × Contractor percentage share

Government profit oil = Profit oil × Government percentage share

Net cash flow = Gross revenues
(aftertax)   – Royalties
– Tangible capital costs
– Intangible capital costs[1]
– Operating costs
+ Investment credits
– Bonuses
– Government profit oil
– Taxes

Taxable income = Gross revenues
– Royalties
– Intangible capital costs[1]
– Operating costs
+ Investment credits
– Government profit oil
– DD&A (including abandonment costs)
– Bonuses[2]

[1]In many systems no distinction is made between operating costs and intangible capital costs. Both are expensed.
[2]Bonuses are not always deductible.

41

SECOND: COST RECOVERY

Before sharing of production, the contractor is allowed to *recover costs* out of net revenues. However, most PSCs will place a limit on cost recovery. For example, in the flow diagram, cost recovery is limited to 40% of gross revenues. If operating costs and DD&A amounted to more than that, the balance would be carried forward and recovered later. From a mechanical point of view, the cost recovery limit is the only true distinction between concessionary systems and PSCs.

THIRD: PROFIT OIL SPLIT

Revenues remaining after royalty and cost recovery are referred to as *profit oil* or *profit gas*. The analog in a concessionary system would be taxable income. The terminology is precise because of the ownership issue. The term *taxable income* implies ownership that does not exist yet under a PSC. The contractor has nothing to tax.

In this example, the contractor's share of profit oil is 40%. The contractor's share of the profit oil is subject to taxation. If this were a service agreement with the contractor's share of revenues equal to 40%, it would be called the *service fee*—not profit oil.

FOURTH: TAXES

The contractor's share of profit oil in this example is taxed at a rate of 40%.

CONTRACTOR TAKE

With cost recovery the contractor's gross share of production comes to 52%. Total profit is $12.00. Considering the 10% royalty, profit oil split, and taxation, the contractor share of profits is $2.40. Contractor take therefore is 20%. The profit margin here looks as though it were 12%. But there may not have been any true profits in the ordinary accounting sense. The cost recovery limit forces some profit sharing under all circumstances where there is production. The important number is the 20% contractor take.

Table 4–1 gives a slightly different perspective on the termi-

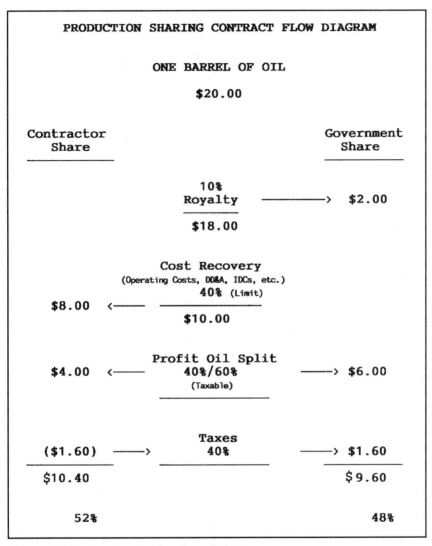

Figure 4–1 Production sharing flow diagram

nology. It focuses on the hierarchy of calculations for contractor cash flow under a PSC.

THE INDONESIAN 85%/15% SPLIT

The most famous government/contractor take statistic is the Indonesian 85%/15% split. There is an important reason for this—Indonesia has no royalty. The calculation is based on two

Table 4-1

Production Sharing Contract Structure
Contractor's Perspective

Terminology	$/bbl	Royalties, Costs, Taxes Sharing
Wellhead price	$20.00	
	− 2.00	10.0% Royalty
Net revenue	18.00	
	0	Local taxes (usually)
Cost recovery elements		
	−4.15	Operating costs
	−1.45	General & administrative costs
	−5.15	Depreciation, depletion & amortization
Total cost recovery	−10.75	
Profit oil	7.25	Sharable oil
Government share (60%)	−4.35	60%/40% Split in favor of Gvt.
Contractor share (40%)	2.90	60%/40% Split in favor of Gvt.
	−1.45	50% Income tax
After-tax net income	$1.45	
	+5.15	Depreciation, depletion & amortization
	−1.25	Tangible capital costs
After-tax cash flow	$5.35	

mechanisms: (1) a profit oil split of 71.1538%/28.8462% in favor of the government and (2) an effective tax rate of 48%. This is the result of a double layer of taxation: 35% income tax and 20% withholding tax levied after income tax.

The effect of this production sharing/tax arrangement in Indonesia is that of an aggregate 85% tax rate. The government share of profit oil could simply be viewed as another layer of taxation.

Because there is no royalty, the division of production is the same regardless of the amount of costs as long as the contractor is able to recover all costs. This is why the 85%/15% split in Indonesia is so well known. It does not change with variations in the level of costs.

Gross production	=	100 MMBBLS
Royalty	=	0%
Costs oil for recovery of costs	=	35 MMBBLS

INDONESIAN PSC ENTITLEMENT CALCULATION

100.00 MM	Gross production
−35.00	Cost recovery
65.00	Total share oil
−46.25	Government share 71.1538%
18.75	**Contractor share 28.8462%**
− 9.00	Aggregate tax rate 48%
9.75	Contractor after-tax share
+35.00	Cost recovery
44.75 MM	**Contractor total financial entitlement**
53.75 MM	**Contractor legal entitlement = (35 + 18.75)**
15.00%	**Contractor take**
	(Contractor after-tax share ÷ Total share oil)
	* See Table 4-2

Table 4-2

Classification of Reserves			
Gross Recoverable Reserves	Contractor 50% Working Interest Share	Contractor Entitlement	Contractor Liftings*
100 MMBBLS	50 MMBBLS	22.375 MMBBLS	22.375 MMBBLS
	Half of 100 MMBBLS	Half of 44.75 MMBBLS from above	
*Ultimately should equal actual entitlements. In any given accounting or reporting period may be more or less than actual entitlement.			

There are other aspects of the Indonesian system that affect the contractor take. These are ignored when the 85%/15% split is mentioned because from the perspective of exploration and development economics, they are not material. They are explained in detail later in this chapter.

RESERVE CLASSIFICATION AND REPORTING

In the previous example, gross production is 100 MMBBLS. Assume that a company holds a 50% working interest in this field. The company's reserve position is summarized in Table 4–2. The gross reserve figures are used when discussing field size thresholds for exploration and development. When discussing the value of reserves in the ground, the gross reserves, or working-interest reserves are used. However, the contractor entitlement is the basis for crude liftings and financial reporting. The entitlement figure is based in part on cost recovery, which depends on oil prices. Therefore, with each reporting period and changing prices, the entitlement reserve figure will change. For example, if oil prices are expected to be lower than those assumed in the previous calculation, the entitlement would be greater. With lower oil prices, the contractor cost oil requirement increases.

In any given accounting or reporting period, the contractor liftings can be different than the actual entitlement because of under- or overliftings due to price changes and various other factors. Because of these differences in reserve classification, it can sometimes be confusing when someone is discussing reserves. Which classification are they talking about?

Most of the discussion in this book centers on the microeconomics of exploration and development of working-interest reserves. Financial reporting, on the other hand, uses reserves associated with contractor entitlement. This number is even more elusive than ordinary reserves estimates. As mentioned previously, entitlement is a function of oil prices. A perfect example of the impact is found in the reserves disclosure segment of the Maxus Energy Corporation 1991 Annual Report [page 47, footnote (b)]:

(b) The 1990 and 1991 changes reflect the impact of the change in the price of crude oil on the barrels to which the Company is entitled under the Indonesian production sharing contracts. The 1990 change due to the impact of increasing prices was a reduction of 20.7 million barrels. Decreasing prices in 1991 resulted in an increase of 25.6 million barrels.

The ownership relationship under a production sharing contract has forced Maxus to report entitlement reserves. Footnote (c) is another matter.

(c) Subsequent to year-end 1991, Maxus signed a letter of understanding to become operator of Block 16 in Ecuador and to increase its ownership interest from 15% to 35%. This will add 36.7 million barrels of new reserves.

Ecuador uses a service contract. Maxus does not have title to these reserves. Maxus could easily be entitled to a share of revenues for cost recovery and a share of profits, but not a share of production. Yet somehow an attempt was made to communicate to shareholders the magnitude of reserves involved. Presumably the 36.7 MMBBLS relate to an equivalent of entitlement to profits and not to working-interest barrels.

The virtue of working-interest barrels is that they are subject primarily to engineering principles and will not change drastically with oil price changes. Working-interest reserves provide common ground for nearly all parties concerned. Financial reporting starts with a cash flow projection of gross or working-interest reserves. Contractor entitlement depends on the level of costs and oil prices. *Entitlement reserves* often amount to around 50% of working-interest reserves. Therefore, the value per barrel from oil company annual and 10-K reports is about double the value of working-interest reserves.

Sample PSC Cash Flow Projection

The cash flow projection in Table 4–3 outlines elements of a simple PSC with a detailed explanation of the calculations involved in arriving at year-by-year cash flow. The PSC consists of:

$$
\begin{array}{rcl}
\text{Royalty} & = & 0\% \\
\text{Cost recovery limit} & = & 60\% \\
\text{DD\&A} & = & \text{5-year straight-line decline (SLD)} \\
\text{Profit oil split} & = & 35\% \text{ for the contractor} \\
\text{Taxes} & = & 40\%
\end{array}
$$

The development costs are all capitalized, and depreciation starts when production begins. The last column (N) is the undiscounted contractor net after-tax cash flow.

The development scenario outlined in this projection is for a 40-MMBBL field with a projected 11-year life. Total capital costs are $60 million, and estimated operating costs during the life of the field are $117 million. Costs represent 22% of gross revenues.

Calculation of Government/Contractor Take

Gross revenues	$799,864	
Total costs	–176,993	
Total profit	622,871	
Government profit oil	–404,865	($622,870 – $218,005)
Government taxes	– 89,002	(includes $3,000 bonuses)
Contractor take	$129,003	($218,005 – $89,002)
Contractor take	21%	($129,003/$622,871)
Government take	79%	
Contractor financial entitlement	305,996	($129,003 + $176,993)
	38%	($305,996/$799,864)

Note: There are some minor rounding differences.

BASIC ELEMENTS

The basic elements of a production sharing system are categorized in Table 4–4. These elements are also found in concessionary systems with the exception of the cost recovery limit and production sharing. Each of the economic elements listed in the table are discussed separately. Noncommercial aspects are covered in Chapter 8.

As this table shows, many aspects of the government/contractor relationship may be negotiated but some are normally determined by legislation. Those elements that are not legislated must be negotiated. Usually, the more aspects that are subject to negotiation, the better. This is true for the government agency responsible for negotiations as well as for the oil companies. Flexibility is required to offset differences between basins, regions, and license areas within a country.

Table 4-3

Sample Production Sharing Contract Cash Flow Projection

Year	OIL Production (MBBLS) (A)	OIL Price ($/bbl) (B)	Gross Revenues ($M) (C)	Intangible Cap Ex ($M) (D)	Tangible Cap Ex ($M) (E)	Operating Expense ($M) (F)	Bonus ($M) (G)	DD&A ($M) (H)	Contractor Cost Oil ($M) (I)	Total Profit Oil ($M) (J)	Contractor Profit Oil ($M) (K)	Tax Loss C/F ($M) (L)	40.0% Income Tax ($M) (M)	Net Cash Flow ($M) (N)
1994	0	$20.00	0	0	10,000	0	2,000	0	0	0	0	0	0	(12,000)
1995	0	20.00	0	0	8,000	0	0	0	0	0	0	2,000	0	(8,000)
1996	0	20.00	0	0	15,000	0	0	0	0	0	0	2,000	0	(15,000)
1997	4,500	20.00	90,000	15,000	10,000	11,500	1,000	8,600	35,100	54,900	19,215	2,000	6,486	10,329
1998	7,000	20.00	140,000	2,000	0	14,000	0	8,600	24,600	115,400	40,390	0	16,156	32,834
1999	5,600	20.00	112,000	0	0	12,600	0	8,600	21,200	90,800	31,780	0	12,712	27,668
2000	4,760	20.00	95,200	0	0	11,760	0	8,600	20,360	74,840	26,194	0	10,478	24,316
2001	4,046	20.00	80,920	0	0	11,046	0	0	19,646	61,274	21,446	0	8,578	21,468
2002	3,439	20.00	68,782	0	0	10,439	0	0	10,439	58,343	20,420	0	8,168	12,252
2003	2,923	20.00	58,465	0	0	9,923	0	0	9,923	48,541	16,990	0	6,796	10,194
2004	2,485	20.00	49,695	0	0	9,485	0	0	9,485	40,210	14,074	0	5,629	8,444
2005	2,087	20.00	41,744	0	0	9,087	0	0	9,087	32,657	11,430	0	4,572	6,858
2006	1,732	20.00	34,647	0	0	8,732	0	0	8,732	25,915	9,070	0	3,628	5,442
2007	1,427	20.00	28,411	0	0	8,421	0	0	8,421	19,990	6,997	0	2,799	4,198
2008	0	20.00	0	0	0	0	0	0	0	0	0	0	0	0
	40,000		799,864	17,000	43,000	116,993	3,000	43,000	176,993	622,870	218,005		86,002	129,003

(A) Production Profile
(B) Crude Price
(C) Gross Revenues = (A) x (B)
(D) Intangible Capital Costs (Expensed) [However, Preproduction costs are often capitalized]
(E) Tangible Capital Costs [Capitalized - see Column (H)]
(F) Operating Expenses (Expensed)
(G) Bonuses are typically not "cost recoverable" but are tax deductible. In this example the tax loss is carried forward until production begins.

(H) Depreciation of Tangible Capital Costs: 5-Year Straight Line Decline
(I) Contractor Cost Oil = (D) + (F) + (H) if (C) is greater than zero: Up to a maximum of 60% of (C)
(J) Total Profit Oil = (C) − (I)
(K) Contractor Profit Oil = (J) x 35%
(L) Tax Loss Carry Forward (See Column H)
(M) Income Tax (40%) = ((K) − (L)) x 40% If (K) − (G) − (L) > (zero) then ((K) − (G) − (L)) x 40% otherwise (zero)
(N) Contractor After-tax Net Cash Flow = (C) − (D) − (E) − (F) − (G) − (J) + (K) − (M)

Table 4-4

Production Sharing Fiscal/Contractual Structure		
	National Legislation	Contract Negotiation
Operational Aspects	• Gvt. participation • Ownership transfer • Arbitration • Insurance	• Work commitment • Relinquishment • Commerciality
Revenue or Production Sharing Elements	• Royalties* • Taxation* • Depreciation rates • Investment credits • Domestic obligation • Ringfencing	• Bonus payments • Cost recovery limits • Production sharing*
*Those features most commonly associated with contractor take.		

The legislative body ordinarily has more authority than the national oil company or oil ministry empowered to negotiate contracts. Therefore, some items are simply not subject to negotiation. If a national oil company has authority to negotiate the profit oil split, then the tax rate is not such an issue. While fiscal elements such as taxes are normally legislated, others are subject to negotiation and are defined in the PSC. Only a contract can set forth such elements as contract area coordinates, work commitment, and duration of phases of the contract. Governments that use the PSC often allow more latitude in negotiations.

WORK COMMITMENT

Work commitments are generally measured in kilometers of seismic data and number of wells. There are some instances where the work commitment may consist only of seismic data acquisition with an option to drill. These are referred to as seismic options. Other contracts have hard, aggressive drilling obligations.

The terms of the work commitment outline penalties for nonperformance.

The work commitment is a critical aspect of international exploration. It embodies most of the risk of petroleum exploration. With most exploration efforts, taxes are never experienced because so many wildcats are dry. There is perhaps only a 10%–15% chance of ever getting beyond the work commitment. Negotiators focus a lot of attention on the work commitment. This is explained in greater detail in Chapter 8.

The main variables that are involved in fiscal design are shown in Illustration 4–1. The designers standing at the fiscal control panel are contemplating a classic production sharing system. To create a concessionary system as far as the arithmetic goes, they would simply turn the cost recovery limit and the contractor profit oil split up to 100%. That would leave a simple royalty/tax system. Not a huge difference.

BONUS PAYMENTS

Cash bonuses are sometimes paid upon finalization of negotiations and contract signing, hence the term *signature bonus*. Although cash payments are most common, the bonus may consist of equipment or technology. Not all PSCs have bonus requirements. Among contracts that have bonus provisions, there are many variations.

Production bonuses are paid when production from a given contract area or field reaches a specified level—usually some multiple of 1,000 BOPD. For example, a contract may require a U.S. $2 million bonus payment to the government when production reaches 20,000 BOPD and another U.S. $2 million bonus if production exceeds 40,000 BOPD. There will be a specified time period such as a month or quarter during which the average production rate must exceed the benchmark level that triggers the bonus payment. Sometimes the production bonuses are paid at startup of production or upon reaching a milestone in cumulative production.

Illustration 4-1 Fiscal design

ROYALTIES

Royalties are a fundamental concept, and the treatment is similar under almost all fiscal systems. While there are some exotic variations on the royalty theme, they are rare. Royalties are taken right off the top of gross revenues. Some systems will allow a *netback* of transportation costs. This occurs when there is a difference between the point of valuation for royalty calculation purposes and the point of sale. Transportation costs from the point of valuation to the point of sale are deducted (netted back).

The concept of a royalty should be foreign to a PSC. This is because of the ownership issue. Many PSCs do not have a royalty. The ones that do range as high as 15%. A PSC royalty is treated just as it would be under a concessionary system. It is the first calculation made. Payment of a royalty implies ownership on the part of the royalty payer. But in a PSC, the contractor has no ownership at this stage. The primary reason that this terminology is used is because of the hierarchy of the arithmetic associated with royalties.

In the Philippines, given sufficient level of Filipino ownership (30% onshore or 15% in deepwater), the government pays the contractor group 7.5% of gross revenues—right off the top. This part of the contractor fee is equivalent to or may be viewed as a *negative royalty*. It is discussed in more detail in Chapter 5.

In New Zealand a hybrid royalty scheme has been proposed. Either a 20% Accounting Profits Royalty (APR) is levied or a 5% Ad Valorem Royalty (AVR), whichever is higher. The AVR is similar to the basic industry royalty levied on gross revenues. The APR, as the first part of the name implies, allows virtually all ordinary accounting deductions in computing the royalty. The APR therefore is not a typical royalty. It behaves more like tax on income.

A specific rate royalty is a fixed amount charged per barrel or per ton. This kind of royalty is relatively rare, but it may also go by another name, such as "export tariff," like that found in the former Soviet Union (FSU). Another type of specific rate royalty is the $1.00/bbl (900 peso) War Tax levy in Colombia. This additional levy was enacted in 1990 and is intended to cover the first six years of production. These are even more regressive than an ordinary royalty.

Another aspect of royalties that contributes to their lack of popularity with industry is that they can cause production to become uneconomic prematurely. This works to the disadvantage of both industry and government.

The royalty scale in Illustration 4–1 ranges from zero to 20%. Anything above 15% is getting excessive. It is inefficient and counterproductive to have royalties too high. A 20% royalty on $18.00 oil is $3.60/bbl. This makes a huge difference with small field developments and marginal production. A marginal field, for example, may require up to 50% of gross revenues over the life of the field for recovery of capital and operating costs. A 20% royalty in that situation would represent 40% of profits. That is way too much. One remedy that has become popular is to scale royalties and other fiscal elements to accommodate marginal situations.

Sliding Scales

A feature found in many petroleum fiscal systems is the sliding scale used for royalties, taxes, and various other items. The most common approach is an incremental sliding scale based on average daily production. The following example shows a sliding scale royalty that steps up from 5% to 15% on 10,000 BOPD tranches of production. If average daily production is 15,000 BOPD, the aggregate effective royalty paid by the contractor is 6.667% (10,000 BOPD at 5% + 5,000 BOPD at 10%).

SAMPLE SLIDING SCALE ROYALTY

Average	Daily Production	Royalty
First Tranche	Up to 10,000 BOPD	5%
Second Tranche	10,001–20,000 BOPD	10%
Third Tranche	Above–20,000 BOPD	15%

Sometimes misconceptions arise when it is assumed that once production exceeds a particular threshold *all* production is subject to the higher royalty rate. Sliding scales do not work that way unless specified in the contract and that is very rare.

Production levels in sliding scale systems must be chosen carefully. If rates are too high, then the system effectively does not have a flexible sliding scale. In some situations tranches of 50,000 BOPD can be too high, or conversely 10,000 BOPD tranches may be too low. The choice is dictated by the anticipated size of discoveries. For a point of reference, most fields worth developing in the international arena will produce from 15%–25% of their reserves in a peak year of production. Generally, the larger the field, the lower the peak percentage rate. Therefore, a 100-MMBBL field might be expected to produce perhaps around 15% of its reserves or 15 MMBBLS in the peak year of production. This is an average daily rate of 41,000 BOPD. If a region is not capable of yielding 100-MMBBL fields, then sliding scale tranches of 50,000 BOPD are useless. There are many variations on the sliding scale theme. Other aspects of the approach are discussed in Chapter 5.

The sequence of calculations that follows the royalty calculation always leads to the recovery of costs. Under the concessionary system these are called *deductions*. Contractual systems use a more descriptive term called *cost recovery*.

COST RECOVERY

Cost recovery is the means by which the contractor recoups costs of exploration, development, and operations out of gross revenues. Most PSCs have a limit to the amount of revenues the contractor may claim for cost recovery but will allow unrecovered costs to be carried forward and recovered in succeeding years. Cost recovery limits or cost recovery *ceilings,* as they are also known, if they exist, typically range from 30%–60%.

Cost recovery is an ancient concept. Even communists are comfortable with it. The ones who put up the capital should *at least* get their investment back. Beyond that, there is wide disagreement. The cost recovery mechanism is one of the most common features of a PSC. It is only slightly different than the cost recovery techniques used in most concessionary systems.

Sometimes the hierarchy of cost recovery can make a difference in cash flow calculations. This is particularly the case if certain cost recovery items are taxable. Cost recovery or *cost oil* normally includes the following items listed in this order:

1. Unrecovered costs carried over from previous years
2. Operating costs
3. Expensed capital costs
4. Current year DD&A
5. Interest on financing (usually with limitations)
6. Investment credit (uplift)
7. Abandonment cost recovery fund

Once the original exploration and development costs are recovered, operating costs comprise the majority of recovered costs. At this stage, cost recovery may range from 15%–30% of revenues.

In most respects, cost recovery is similar to deductions in calculating taxable income under a concessionary system. The profit oil share taken by the government could be viewed as a first layer of taxation. However, the terminology is specific and harkens back to the ownership issue. A contractor under a PSC does not own the production, and therefore at the point of cost recovery has no taxable revenues against which to apply deductions. The government reimburses the contractor for costs through the cost recovery mechanism and then shares a portion of the remaining production or revenues with the contractor. It is only at that point then that taxation becomes an issue.

Exceptions

While the cost recovery treatment is common in the universe of PSCs and service agreements, there are exceptions to every rule. Some contracts have no limit on cost recovery. The second generation Indonesian PSC had no such limit. From a mechanical point of view, it was not a PSC. And there are other PSCs that have no limit on cost recovery. Some PSCs have *no* cost recovery!

The 1971 and 1978 Peruvian model contracts made no allowance for cost recovery prior to the profit oil split. The government simply granted the contractor a share of production which ranged from 44%–50%, depending upon the contract area. The contractor also paid a tax on net income. This type of sharing arrangement was also found in the 1975 Trinidadian offshore contract with Mobil. There was no provision for cost recovery; production was simply divided 60%/40% in favor of the government up to 50,000 BOPD and 65%/35% from 50,000 to 100,000 BOPD. This unusual feature is viewed by some as a royalty. The government took its share right off the top, regardless of deductions or cost recovery requirements.

Another exception is where excess cost oil goes directly to the government. Although this feature is quite rare it is noteworthy. This provision was proposed by Egypt during the Nile Delta licensing round in 1989. It is also found in the Syrian PSC, which has a 25% cost recovery ceiling. Figure 4–2 illustrates this variation on the cost recovery theme. The cost recovery ceiling in this example is 40%, but the contractor recovered eligible costs out of 21% of gross revenues in this example. The remaining 19% then went directly to the government. It was not subject to the profit oil split.

Tangible vs. Intangible Capital Costs

Sometimes a distinction is made between depreciation of fixed capital assets and amortization of intangible capital costs. Under some concession agreements, intangible exploration and development costs are not amortized. They are expensed in the year they are incurred and treated as ordinary operating expenses. Those rare cases where intangible capital costs are written off immediately can be an important financial incentive. Most systems will force intangible costs to be amortized. Therefore, recovery of these costs takes longer, with more revenues subject to taxation in the early stages of production.

Many accounting conventions around the world make little or no distinction between intangible capital costs and operating costs that are expensed. The exception is with intangible exploration and development costs prior to production. These preproduction

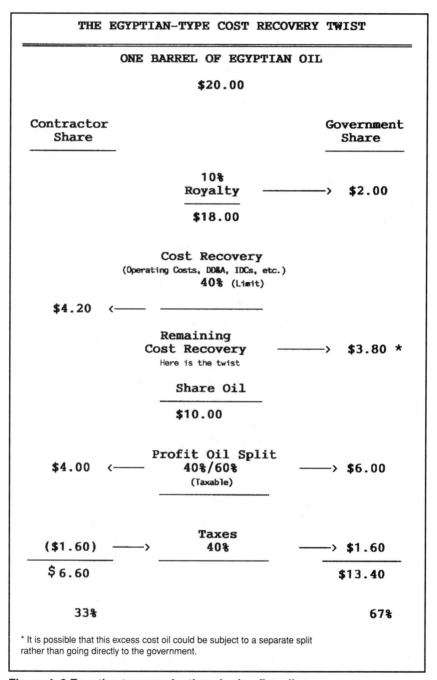

Figure 4–2 Egyptian-type production sharing flow diagram

costs under most systems are amortized beginning with production startup. Depreciation rates are the other primary limitation to the rate of recovery of capital costs. This is true for all fiscal systems that require capitalization of costs.

Interest Cost Recovery

Sometimes interest expense is allowed as a deduction. This can include interest during construction or a rate based on cumulative unrecovered capital. Under a PSC this is referred to as interest cost recovery. Some systems limit the amount of interest expense by using a theoretical capitalization structure such as a maximum 70% debt. In Papua New Guinea, the government limits the interest deduction with a *capitalization restriction* based on a 2/1 debt/equity ratio. Some systems will limit the interest rate itself, regardless of the actual rate of interest incurred.

China allows recovery of a 9% annual *deemed* interest rate on development costs. This rate is compounded annually and recoverable through cost recovery along with development costs.

This concept can take on added dimension and begin to look like a rate-of-return feature that is being found in more and more fiscal systems. It is discussed further in Chapter 5, which deals with ROR contracts. It is covered further later in this chapter in the discussion of the Indonesian PSC.

General & Administrative Costs (G&A)

Many systems allow the contractor to recover some home office administrative and overhead expenses. Nonoperators are normally not allowed to recover such costs. Contractors in Indonesia are limited to 2% of gross revenues for G&A cost recovery.

The 1989 Myanmar model contract had a sliding scale allowance for G&A based on total petroleum costs each year:

For the first U.S. $5 million: 4%

For the next U.S. $3 million: 2%

For the next U.S. $4 million: 1%

Over U.S. $12 million: 0.5%

In China an annual overhead charge is allowed for offshore exploration at a rate of 5% on the first $5 million per year, dropping down to 1% for costs above $25 million. Development overhead charges are allowed at a rate of 2.5% on the first $5 million, dropping down to 0.25% on costs exceeding $25 million. Overhead for operating costs was allowed at a flat rate of 1.8%.

UNRECOVERED COSTS CARRIED FORWARD

Most unrecovered costs are carried forward and are available for recovery in subsequent periods. The same is true of unused deductions. The term *sunk cost* is applied to past costs that have not been recovered. There are four classes of sunk cost:

- Tax Loss Carry Forward (TLCF)
- Unrecovered Depreciation Balance
- Unrecovered Amortization Balance
- Cost Recovery Carry Forward

These items are typically held in abeyance prior to the beginning of production. Many PSCs do not allow preproduction costs to begin depreciation or amortization prior to the beginning of production, so there is no TLCF. Bonus payments, though, may create a TLCF.

Exploration sunk costs can have a significant impact on field development economics, and they can strongly affect the development decision. The importance of sunk costs and development feasibility centers on an important concept called *commerciality*. This issue is discussed in detail later in this chapter.

The financial impact of a sunk cost position on the development decision can be easily determined with discounted cash flow analysis. The field development cash flow projection should be run once with sunk costs and once without. The difference in present value between the two cash flow projections is the present value of the sunk cost position. For example, if a company has $20 million in sunk costs and is contemplating the develop-

Table 4-5

COST RECOVERY SPECTRUM
RANGE OF COST RECOVERY LIMITS

20%	40%	60%	80%	100%

[1]	Cruel & Unusual	Low End Typical	Upper End Rare[2]	More Rare [3]	[4]

[1]No examples in this author's experience.
[2]Cost recovery limits of 40%–60% probably encompass over 75% of the fiscal systems that have a limit.
[3]Indonesia had no limit on cost recovery for many years and now with the 20% "First Tranche Petroleum" has the equivalent of an 80% cost recovery limit.
[4]Concessionary systems usually have no limit on cost recovery.

ment of a discovery, the present value of the sunk costs could be on the order of $10–$15 million. This would depend on whether or not the costs could be expensed or if they must be capitalized. It also depends on restrictions on cost recovery—primarily the cost recovery limit.

Table 4-5 shows the range of cost recovery limits found in the universe of PSCs.

With marginal field development a low cost recovery limit has a big impact. For example, a cost recovery limit of 50% or less in a marginal situation can have effectively the same impact on project NPV and IRR as a 5–10 percentage point reduction in contractor take. However, with large profitable fields, even the lower levels of cost recovery are much less an issue. The problem for fiscal design and negotiations lies in estimating the range of field sizes likely to be found and structuring provisions that can handle them.

Abandonment Costs

The issue of ownership adds an interesting flavor to the concept of abandonment liability. Under most PSCs the contractor cedes ownership rights to the government for equipment, platforms, pipelines, and facilities upon commissioning or startup. The government as owner is theoretically responsible for the cost of abandonment. In fact, governments are responsible, and they do pay for abandonment costs, but they do it indirectly. The government ultimately pays for abandonment just as it paid for drilling and development. Abandonment costs are recovered through cost recovery just as the costs of exploration, development, and operations are. Anticipated cost of abandonment is accumulated through a sinking fund that matures at the time of abandonment. The costs are recovered prior to abandonment so that funds are available when needed.

Profit Oil Split and Taxation

Profit oil is split between the contractor and the government, according to the terms of the PSC. Sometimes it is negotiable. The contractor's share of profit oil is usually subject to taxation. Published government or contractor take figures refer to the after-tax split.

Explorers focus on geopotential and how that potential balances with fiscal terms and the cost of doing business. When evaluating fiscal terms the focus is on division of profits—the government/contractor take. Figure 4–3 illustrates the effective tradeoff. Geopotential, costs, infrastructure, political stability, and other key factors that influence business decisions are weighed against contractor take. The split in most countries ranges from just under 15% to over 55% for the contractor. Beyond these extremes are the exceptions that are becoming more and more rare.

Governments are not totally responsible for determining the appropriate division of profits and contract terms. Oil companies help define what the market can bear—as shown in Illustration 4–2. This situation happens all the time.

COMMERCIALITY

An important aspect of international exploration is the issue of *commerciality*. It deals with who determines whether or not a

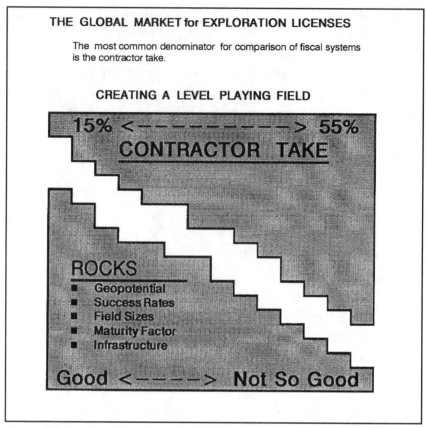

Figure 4–3 Creating a level playing field

discovery is economically feasible and should be developed. It is a sensitive issue. This is because there are often situations where accumulated exploration expenditures are so substantial that by the time a discovery is made, these sunk costs have a huge economic impact on the development decision. From the perspective of the contractor, these sunk costs upon development will flow through cost recovery (or will be used as deductions), and they can represent considerable value. But they represent a liability, or a cost, as far as the government is concerned. If cost recovery is too great, then the government may end up with only a small percentage of the gross production, depending upon the contractual/fiscal structure.

Illustration 4–2 Realities of the marketplace

Some regives will simply allow the contractor to decide whether or not to commence development operations. Other systems have a *commerciality requirement*. This requirement essentially places the burden of proof on the contractor as to whether or not development of a discovery is economically beneficial for *both* the contractor and the government. The benchmark for obtaining commercial status for a discovery is usually a predetermined percentage of gross take for the government. Under many commerciality clauses, a discovery cannot be developed unless it is granted commercial status by the host government. The grant of commercial status marks the end of the exploration phase and the beginning of the development phase of a contract.

In Colombia the commerciality issue is complicated by the government's, 50%-carry through the exploration and delineation phase. Government participation effectively begins at the point that commercial status is granted. If the government does not agree that the discovery can justify development, the contractor may still go forward. In that case, the government back-in does not take effect until the contractor recoups 200% of the investment. Then the government backs in. Up to that point, the government receives a 20% royalty.

The issue is particularly critical with progressive regimes where government take is based more on profitability than gross revenues. If a marginal field is developed under a progressive regime then the government share of revenues could be both small and substantially delayed. This creates an important consideration that is critical in fiscal design. There is a trade-off. The systems with significant limits on the contractor's access to gross revenues have little need for a commerciality requirement. But the countries that have no cost recovery limits and low royalties often protect themselves with a commerciality clause. The discussion of commerciality under the Indonesian PSC later in this chapter illustrates the trade-off.

GOVERNMENT PARTICIPATION

Many systems provide an option for the national oil company to participate in development projects. Under most government participation arrangements, the contractor bears the cost and risk of exploration. If there is a discovery, the government backs-in for a percentage. In other words, the government is *carried through exploration*. This is fairly common and automatically assumed whenever some percentage of government participation is quoted.

Both the Indonesian and Malaysian PSCs have government participation clauses, but Indonesia rarely exercises its option to participate.

The key aspects of government participation are:

- What percentage participation?
  (Most range from 10%–51%.)
- When does the government back in?
  (Usually once a discovery has been made.)
- How much participation in management?
  (Large range of degree of participation.)
- What costs will the government bear?
  (Usually only their prorated share of development costs.)
- How does government fund its portion of costs?
  (Often out of production.)

The financial effect of a government partner is similar to that of any working-interest partner with a few *large* exceptions. First, the government is usually *carried* through the exploration phase and may or may not reimburse the contractor for past exploration costs. Second, the government's contribution to capital and operating costs is normally paid out of production. Finally, the government is seldom a silent partner.

In Colombia the government has the right to take up to 50% working interest and will reimburse the contractor up to 50% of any *successful* exploratory wells. In China the government participation is 51%. This usually defines the upper limit of direct government involvement.

Contractors prefer no government participation. This is not totally selfish, but stems from a desire for efficiency as well as economy. Joint operations of any sort, especially between diverse cultures, can have a negative impact on operational efficiency. This is particularly true when the interests of government and an oil company can be so polarized.

INVESTMENT CREDITS AND UPLIFTS

Some systems have incentives, such as investment credits or uplifts. Uplifts and investment credits are two names for the same basic concept. An *uplift* allows the contractor to recover an additional percentage of capital costs through cost recovery. It works the same way in a concessionary system. For example, an uplift of 20% on capital expenditures of $100 million would allow the contractor to recover $120 million. Uplifts can create incentives for the industry, and there are a number of different designs. Uplifts are a key aspect of rate-of-return contracts which are discussed in Chapter 5.

DOMESTIC OBLIGATION

Many contracts have provisions that address the domestic crude oil or natural gas requirements of the host nation. These provisions are often referred to as the domestic supply requirement or domestic market obligation (DMO). Usually they specify that a certain percentage of the contractor's profit oil be sold to the government. The sales price to the government is usually at a discount to world prices. The government may also pay for the domestic crude in local currency at a predetermined exchange rate. Revenues from sale of domestic crude are normally taxable.

RINGFENCING

The issue of recovery or deductibility of costs is further defined by the revenue base from which costs can be deducted. Ordinarily all costs associated with a given block or license must be recovered from revenues generated within that block. The block is *ringfenced*. This element of a system can have a huge impact on the recovery of costs of exploration and development. Indonesia requires each contract to be administered by a separate new company. This restricts consolidation and effectively erects a ringfence around each license area.

Some countries will allow certain classes of costs associated with a given field or license to be recovered from revenues from another field or license. India allows exploration costs from one area to be recovered out of revenues from another, but development costs must be recovered from the license in which those costs were incurred.

From the government perspective, any consolidation or allowance for costs to cross a ringfence means that the government may in effect subsidize unsuccessful operations. This is not a popular direction for governments because of the risky nature of exploration. However, allowing exploration costs to *cross the fence* can be a strong financial incentive for the industry.

The importance of risk dollars has already been demonstrated. If a country with an effective tax burden of 50% allowed exploration costs to be deducted across license boundaries, then the industry could be drilling with 50¢ dollars. It would cut the risk in half. From the perspective of the development engineer, this cut would have little meaning unless development and operating costs are also allowed to cross. Dropping or loosening the ringfence can provide strong incentives, especially to companies that have existing production and are paying taxes.

In the late 1980s and early 1990s, exploration in the U.K. sector of the North Sea reached record levels. This is because the government allowed exploration costs to cross the ringfence as deductions against the 75% PRT tax on older fields. This created a huge exploration incentive for any company paying PRT taxes. And exploration drilling shot up. Companies were purchasing what came to be known as *Forties Units.* These were .25% working-interest shares in the British Petroleum-operated Forties field, which during 1992 produced more than 160,000 BOPD. Some of the larger companies had substantial unused tax cover, and smaller companies did not have enough. They could buy a couple of Forties Units and take advantage of the exploration relief provided by the hole in the ringfence. The dynamics of ringfencing can be spectacular. In 1994 the U.K. government abandoned the cross-fence PRT deduction, and exploration dropped off dramatically.

REINVESTMENT OBLIGATIONS

Some contracts require the contractor to set aside a specified percentage of income for further exploratory work within a license.

In France the level of taxation was effectively reduced when the company reinvested a certain portion of income. This approach is not as harsh as a firm obligation. The objective, of course, is to get companies to spend more in-country and repatriate less of their profits. Reinvestment obligations or reinvestment incentives are fairly rare.

TAX AND ROYALTY HOLIDAYS

Governments can enact legislation or issue decrees that are designed to attract additional investment. Tax or royalty holidays are often used for this purpose. These specify that for a given holiday period, royalty or taxes are not payable.

Myanmar built into their PSC for the 1989–90 licensing round a three-year holiday on their 30% income tax. The start of the holiday begins with the start of production. Therefore, the first three years of production from licenses issued at that time will not be burdened with corporate income tax.

When reviewing fiscal terms that have time limitations, the starting point is as important as the time period. In some instances the holiday begins on the effective date of the contract, or on a specific calendar date. In other cases the holiday may begin with production startup.

Holidays can make quick-look analysis a bit difficult. However, there are some basic rules. A tax holiday has no impact if no discovery is made or if there is no production. If there is production, the effect of a holiday is twofold. A reduction in levies, taxes, or royalties will always be beneficial to the petroleum industry, and the holiday almost always comes at the early stage of production when it does the most good in terms of present value.

A production profile is shown in Table 4–6. For convenience, a 100-MMBBL field is used—the annual production, therefore, in millions of barrels (MMBBLS) also represents the percentage of production per year. In the first year, 16.5 MMBBLS are produced (16.5% of the production), and production declines exponentially at a rate of 15% per year thereafter.

In this example, by the end of the fourth year, over 50% of the reserves have been produced. However, in terms of present value discounted at 15%, over 70% of the reserves have been produced. A simple three-year holiday would cover over 60% of the reserves in terms of present value. This production profile is relatively conservative. Many Southeast Asia fields have produced at much faster rates, with many producing as much as 25% of their ultimate recoverable reserves in the first year.

THE INDONESIAN PSC

Understanding the current Indonesian PSC and how it evolved leads to good general understanding of PSCs and the foundation of all contractual systems. There are a number of reasons for this, the most obvious one being that the first PSCs ever were signed there in 1966. Since then Indonesia has been one of the most active countries in Southeast Asia with nearly half of all contractors/licenses in the region. There were over 50 operating companies in Indonesia, and over 100 PSCs had been signed there by 1994. Because so many companies have participated in Indonesia, it is one of the best-known systems in the world. Most negotiators have worked their way through more than one vintage of the Indonesian PSC. It will continue to function as a model for all others.

The sharing of production is the heart and soul of a PSC. The best way to evaluate a PSC is to begin with how the production is shared. The contractor's share is often referred to as the *contractor entitlement*. The contractor entitlement in Indonesia is calculated as follows:

Contractor
Entitlement = Cost recovery
+ Investment credit
+ Contractor equity share (profit oil)
− Domestic market requirement (adjustment)
− Government tax entitlement

Table 4-6

Present Value Profile of Oil Production

Year	Production Profile (MMBBLS)	Cumulative Percentage Produced	Mid-Year Discount Factor 15%	Discounted Production Profile (MMBBLS)	Discounted Percentage Annual	Discounted Percentage Cumulative
1	16.5	16.5%	.933	15.4	26.4%	26.4%
2	14.0	30.5	.811	11.4	19.5	45.9
3	11.9	42.4	.705	8.4	14.4	60.3
4	10.1	52.5	.613	6.2	10.6	70.9
5	8.6	61.1	.533	4.6	7.9	78.8
6	7.3	68.4	.464	3.4	5.8	84.6
7	6.2	74.6	.403	2.5	4.3	88.9
8	5.3	79.9	.351	1.9	3.2	92.1
9	4.5	84.4	.305	1.4	2.3	94.4
10	3.8	88.2	.265	1.0	1.7	96.2
11–15	11.8	100.0	.188 *	2.2	3.9	100.0
	100.0			58.4	100.0%	

*Aggregate discount rate for last five years of production

The basic features common to most PSCs are cost recovery, profit oil, and taxes. The other features of the Indonesian system are specific to Indonesia and are explained later in this chapter. Each aspect of the Indonesian PSC is described in detail, starting with the one aspect that is characteristic because of its absence. Indonesia has no royalty.

NO ROYALTY

The Indonesian PSCs are characterized in part by the lack of a royalty. However, some people refer to the *first tranche petroleum* (FTP) as the equivalent of a royalty. Others view it as more of a cost recovery limit. Both aspects of the FTP are discussed later.

COST RECOVERY

Cost recovery limits (which to a large extent comprise the only mechanical difference between concessionary systems and PSCs) have changed dramatically through the years in Indonesia. The first generation contracts of the 1960s had a 40% limit. The second generation contracts after 1976 did away with the cost recovery limit.

Elements that make up cost recovery are normally recovered on a first-in, first-out basis. Any costs carried forward from prior years are recovered first. The order is as follows:

1. Amortization of noncapital carryforward
2. Depreciation of capital carryforward
3. Unrecovered prior year costs
4. Current year noncapital costs (operating costs)
5. Current year depreciation of capital costs
6. Investment credit

FIRST TRANCHE PETROLEUM

With the fourth-generation contracts outlined in the 1988–89 incentive packages, a new contract feature was introduced. It is called *first tranche petroleum* (FTP). The first tranche petroleum element requires that 20% of production be shared 71.1538%/28.8462% in favor of the government before cost recovery. The contractor's share is taxed.

The FTP is viewed by some as a 14.23% royalty because that is the government before-tax share of the 20% first tranche. However, the contractor share of first tranche petroleum is taxed at the effective rate of 48%. The result is that 3% of gross production goes to the contractor and 17% goes to the government. The remaining 80% of production is available for cost recovery. Hence the FTP works *exactly* like a cost recovery limit. The strongest argument in favor of viewing the FTP as a cost recovery limit instead of a royalty is that it is not regressive like a royalty.

In older Indonesian PSCs, the contractors had to demonstrate that the Indonesian government would ultimately receive a minimum of 49% of the total revenue over the life of a field in order to be granted the *commercial status* required for field development. With the arrival of the 1988–89 Indonesian contracts, this feature of the Indonesian PSC was being simplified, and in some cases it was eliminated altogether. Contracts signed after August 1988 included FTP which eliminated the minimum total revenue requirement for commerciality.

INVESTMENT CREDITS AND UPLIFTS

The Indonesian contracts have allowances for investment credits (ICs) and uplifts. The difference between the two is that the uplift applies to all capital costs, and the IC does not. The IC and the uplift are otherwise similar. The IC applies only to facilities such as platforms, pipelines, and processing equipment. This excludes drilling costs and completion costs.

In Indonesia the investment credit is immediately recoverable and need not be depreciated like the costs which justified the credit. It can also be carried forward.

Investment credits reduce the ultimate profit oil split for both the contractor and the host government. The investment credit benefits are also taxable. The economic impact of the 17%–20% investment credits in Indonesia are almost negligible from the explorationist point of view. But in the joint operating agreements administered by joint operating bodies (JOA/JOB), the uplift is 110% and can have quite an impact.

Because investment credits are taxable, the *order* of cost recovery can be important. In Indonesia the ICs could be carried forward two years, but they were last in the priority of cost recovery. A provision was allowed to put the IC first if it appeared that its carry-forward eligibility would expire. Furthermore, the ICs in Indonesia had other stipulations. They were available only for oil operations, not gas. The credits were also contingent upon Indonesia ultimately getting a 49% share of the production from a field.

The investment credits and uplifts are cost recoverable but not deductible for calculation of income tax. The opposite is true for bonuses, which are not cost recoverable, but they are tax deductible.

INDONESIAN DMO ADJUSTMENT

The Indonesian DMO requires the contractor to sell 25% of the contractor's share oil to Pertamina. The computation for the share oil is based upon the contractor pretax profit oil share of 28.8462%. After 60 months of production from a given field, the price the contractor receives for the DMO crude is 10% of the realized price.

In Indonesia the DMO requirement is based upon total liftings:

Total lifting × 28.8462%   = Contractor share oil
Total lifting × 28.8462% × 25% = DMO (post-1984 contracts)

The obligation is referred to as the *DMO adjustment*. The adjustment is based upon the quantity of DMO crude oil and the difference between the DMO price and actual realized prices.

The DMO adjustment is based upon share oil as previously defined where contractor profit oil is based on the government/contractor equity split following cost recovery. Therefore, the DMO adjustment could theoretically exceed the contractor's share of profit oil. However, the DMO adjustment is defined as the lesser of the contractor profit oil or the DMO adjustment. A sample calculation follows. Assume that during a one-year period, the contractor produced 1 MMBBLS at $20.00/bbl in a given field after 60 months of production:

SAMPLE INDONESIAN DMO ADJUSTMENT CALCULATION

DMO Adjust **= Gross Production × Price Differential × DMO % × Share Oil %**
DMO Adjust = 1 MMBLS × ($20.00 – $2.00) × 25% × 28.8462%
            = $1,298,079   or   64,904 bbl

NEW OIL VS. OLD OIL

In Indonesia, when discussing the domestic market obligation (DMO), the terms *new oil* and *old oil* are used. The first 60 months of production from a field is new production, and the contractor receives market price for the DMO crude. After that the production is referred to as old oil, and the DMO crude sells to the government for 10% of market price. Older contracts pegged the DMO price at 20¢/bbl.

Some of the early PSCs in Indonesia are either expiring or being renegotiated. With renegotiation of these earlier contracts, some of the old fields are still receiving and will continue to receive only 20¢/bbl for old DMO crude. But new fields discovered after a certain date (depending on the contract) will receive 10% of the market price for DMO crude after 60 months of production. Therefore, some production streams from these old contract areas will consist of three classes of crude oil, each receiving a different price:

	New Oil	Old Oil Old Fields	Old Oil New Fields
DMO Crude price =	Market price	20¢/bbl	10% of Market price

Consider the 5-year holiday for the Indonesian DMO. In the sample field profile in Table 4–6, nearly 80% of the reserves are produced in terms of present value by the time the low-price phase of the DMO kicks in. This is an important consideration. The DMO can sound quite harsh, but with a 5-year holiday, it has little meaning as far as exploration economics are concerned.

INDONESIAN CRUDE PRICES

Production is shared in kind (barrels), and it is necessary to determine a price to convert oil to dollars in order to calculate cost recovery, taxes, and internal transfers. Terminology and methodology have changed over the years, but the current method uses the Indonesian Crude Price (ICP), which has been in effect since April 1989. The ICP is determined by the government monthly, based on a moving average spot price of a basket of five internationally traded crudes:

- Indonesian   (Minas)
- Malaysian   (Tapis)
- Australian   (Gippsland)
- UAE   (Dubai)
- Oman

Under the ICP formula, the price of Cinta crude for April delivery, for example, is set at the average of the basket of crudes during the last 15 days of March and the first 15 days of April, plus or minus the difference between the rolling average price of Cinta crude and the basket of crudes during the past 52 weeks. For instance, assume that the average spot price for the basket of crudes for the last 15 days of May and first 15 days of June is $15/bbl. Assume too that Cinta crude during the past 52 weeks has sold at an average price of $1.25/bbl less than the basket of crudes. The ICP price of Cinta crude for June delivery would be calculated as:

Basket average spot price	$15.00
Average 52-week adjustment for Cinta crude	−1.25
Cinta ICP for June delivery	$13.75

Tax calculations are based on ICP, and cash flow is based upon actual realized prices. For most economic modeling, particularly for full-cycle economics and exploration risk analysis, there is no distinction made between estimated market prices and ICP. In the long run, ICP and realized prices will average out any differences.

BONUSES

Bonuses are negotiated for each contract and consist of signature or signing bonuses as well as production bonuses. In the past the Indonesian bonuses payments have been relatively modest. Bonuses are not recoverable through cost recovery, but they are deductible against income and withholding taxes.

EVOLUTION OF THE INDONESIAN CONTRACT

The Mining Law of 1960, Law No. 44 Concerning Oil and Gas Mining, clarified the status of foreign oil companies as *contractors*. This was founded on the 1945 Constitution Article 33, which placed the nation's natural wealth within the jurisdiction of the state. This set the stage for development of the Indonesian PSC in all its variations. Because of the importance of the Indonesian contract its evolution is outlined here.

CONTRACTS OF WORK

The predecessors of the PSC are the early *contracts of work,* a fairly outmoded term that is nearly synonymous with PSC. The early contracts in Indonesia were referred to as such but had virtually all the elements of a PSC or a risk service agreement. The modern PSC in Indonesia is much more complex, but this is where it began.

The cash flow projection in Table 4–7 outlines the basic elements of a fourth generation Indonesian PSC with a detailed explanation of the calculations involved in arriving at year-by-year cash flow.

INDONESIAN CONTRACT EVOLUTION

1954 Original model of future contracts signed with Stanvac

1963 Contracts of work signed by Stanvac, Shell, and CalTex
- 60%/40% Split after Cost Recovery

1966 First Generation PSC—Basic PSC Signed with IIAPCO
- 40% Cost recovery limit
- 65%/35% Split inclusive of taxes in favor of government
  67.5%/32.5% Above 75,000 BOPD
- 25% Domestic Market Obligation (DMO) full price for first 5 years of production, 20¢/bbl thereafter.

1976 Second Generation PSC
- 100% Cost recovery (no limit)
  - 10-year amortization of noncapital costs (SLD)
  - 14-year depreciation of capital costs (DDB)
- 85%/15% Split inclusive of taxes in favor of government
- 20% Investment credit
- 25% DMO 20¢/bbl after 60 months

1978 Changes: Decree 267
- Taxes became payable: **Effective 56% tax**
  - 45% Tax on net income
  - 20% Tax on distributable income after-tax (withholding)
- 85%/15% Split (Oil) determined on after-tax basis
  Oil 65.9091%/34.0909% in favor of the government
- 70%/30% Split (Gas) determined on after-tax basis
  Gas 31.8181%/68.1818% in favor of the contractor

1984 Third Generation PSC—New Tax Laws—Decree 458
- 17% Investment Credit
- **Effective 48% tax**
  - 35% Tax on net income

- 20% Tax on distributable income after-tax (withholding)
- Depreciation: Oil 7-year DDB (switching to SLD in year 5)
  Gas 14-yr DDB (switching to SLD)
- 85%/15% Equity split (Oil)
  Oil 71.1538%/28.8462% in favor of the government
- 70%/30%–65%/35% Equity split (Gas)
  Gas 57.6923%–67.3077% Contractor pretax share

1988–1989 Changes—Fourth Generation PSC—Incentive Packages

- First Tranche Petroleum 20%
- Depreciation: Oil, 5-year DB; Gas, 8-year DB
- Commerciality requirement excluded
- 17% Investment credit
- DMO priced at 10% of export price
- Equity split in frontier areas 80%/20% up to 50,000 BOPD also for marginal fields with less than 10,000 BOPD
- Deepwater investment uplift 110% for Oil, 55% for Gas

1994 Eastern Frontier Incentive Package

- Equity split in Eastern Indonesia 65%/35%

JOINT OPERATING AGREEMENTS/JOINT OPERATING BODY (JOA/JOB)

In the 1970s Pertamina introduced a joint venture arrangement where the contractor participates as a 50%/50% partner with the government in areas previously under Pertamina's control. The first such contract was signed in 1977, and many have been signed since then. Typically the contractor matches previous expenditures for the area or funds 100% of future operations for a given period such as two to three years, whichever is greater. After that, the contractor and Pertamina split exploration, development, and operating costs 50%/50%. Oil production is also shared on a 50%/50% basis. The contractor's 50% share of the

Table 4–7

Sample Indonesian Cash Flow Projection
4th Generation PSC   Post 1988–1989 Incentive Packages

Year	OIL Production (MBBLS) (A)	OIL Price ($/BBL) (B)	Gross Revenues ($M) (C)	Intangible Cap Ex ($M) (D)	Tangible Cap Ex ($M) (E)	Operating Expense ($M) (F)	Bonus ($M) (G)	DD&A ($M) (H)	Investment Credit ($M) (I)	Contractor Cost Oil ($M) (J)	Total Profit Oil ($M) (K)	Contractor Profit Oil ($M) (L)	DMO Adjust ($M) (M)	Income Tax ($M) (N)	Net Cash Flow ($M) (O)
1994	0	$20.00	0	0	10,000	0	2,000	0	0	0	0	0	0	0	(12,000)
1995	0	20.00	0	0	8,000	0	0	0	0	0	0	0	0	0	(8,000)
1996	0	20.00	0	15,000	15,000	0	1,000	0	3,655	0	0	0	0	0	(15,000)
1997	4,500	20.00	90,000	2,000	10,000	11,500	0	12,286	0	42,441	47,559	13,719	0	6,900	11,760
1998	7,000	20.00	140,000	0	0	14,000	0	8,776	0	24,776	115,224	33,238	0	15,954	26,059
1999	5,600	20.00	112,000	0	0	12,600	0	6,268	0	18,868	93,132	26,865	0	12,895	20,238
2000	4,760	20.00	95,200	0	0	11,760	0	4,477	0	16,237	78,963	22,778	0	10,933	16,322
2001	4,046	20.00	80,920	0	0	11,046	0	3,731	0	14,777	66,143	19,080	0	9,158	13,653
2002	3,439	20.00	68,782	0	0	10,439	0	3,731	0	14,170	54,612	15,753	2,232	6,490	10,762
2003	2,923	20.00	58,465	0	0	9,923	0	3,731	0	13,654	44,810	12,926	3,795	4,383	8,479
2004	2,485	20.00	49,695	0	0	9,485	0	0	0	9,485	40,210	11,599	3,225	4,019	4,354
2005	2,087	20.00	41,744	0	0	9,087	0	0	0	9,087	32,657	9,420	2,709	3,221	3,490
2006	1,732	20.00	34,647	0	0	8,732	0	0	0	8,732	25,915	7,475	2,249	2,509	2,718
2007	1,427	20.00	28,411	0	0	8,421	0	0	0	8,421	19,990	5,766	1,844	1,883	2,040
2008	0	20.00	0	0	0	0	0	0	0	0	0	0	0	0	0
	40,000		799,864	17,000	43,000	116,993	3,000	43,000	3,655	180,648	619,215	178,620	16,054	78,346	84,875

(A) Production Profile (New Oil becomes Old Oil after 60 months production) Startup July 1, 1997.
(B) Crude Price [No distinction made here between Realized Price and Indonesian Crude Price (ICP): Taxes based on ICP – Cash Flow based on Realized Price]
(C) Gross Revenues = (A) * (B)
(D) Intangible Capital Costs (Expensed) [However, Preproduction costs are usually capitalized]
(E) Tangible Capital Costs [Capitalized – see Column (H)]
(F) Operating Expenses (Expensed)
(G) Bonuses typically not "cost recoverable" but are tax deductible [The opposite is true of the Investment Credit (I)] $2000 bonus carried forward to 1997.
(H) Depreciation of Tangible Capital Costs: 7 year Double Declining Balance with balance written off in year 5 (3 year SLD)
(I) Investment Credit: Assuming 50% of Tangible Costs (E) are "Eligible" = (E) * 17% * 50% (Flows through cost recovery but not tax deductible)
(J) Contractor Cost Oil = (D) + (F) + (H) + (I) : Up to a maximum of 80% of (C) if (C) is > (zero)
(K) Total Profit Oil = (C) – (J)
(L) Contractor Profit Oil = (K) * 28.8462%
(M) DMO Adjustment: After 60 Months Production – Difference between Market Price and 10% of Market Price for "Share Oil" = (C) * 25% * 28.8462% * 90%
(N) Indonesian effective tax rate resulting from two-tiered tax: Income Tax (35%) and Dividend Tax (20%) = [(L)+(I)–(G)–(M)] * 48%
(O) Contractor After-Tax Net Cash Flow = (C) – (D) – (E) – (F) – (G) – (K) + (L) + (I) – (M) – (N)

production then is subject to the standard PSC terms, i.e., the 85%/15% split, and the FTP, DMO, ICs, etc. Some people view this as an effective 92.5%/7.5% split, but this is a rather extreme view that ignores the government sharing of costs.

Pertamina has the option of paying its share of costs or having the contractor pay on Pertamina's behalf. If Pertamina elects to have the contractor pay its share of costs, these costs are subject to a 50% uplift that flows to the contractor through cost recovery.

The JOA type of joint venture is administered by a Joint Operating Body (JOB) comprised of three representatives each from the contractor and Pertamina.

SENSITIVITY ANALYSIS OF THE INDONESIAN PSC

The complexity of the Indonesian PSC provides a good example for evaluating the economic sensitivity of various contract elements. Cash flow analysis was performed on a 50-MMBBL field development scenario. The basic assumptions are outlined in Table 4–8.

In order to gauge the effect of the DMO and the investment credit, hypothetical scenarios were developed. Three cash flow projections excluded the DMO and the IC. These scenarios yielded pure splits between the government and the contractor of 84%/16%, 85%/15%, and 87%/13% respectively. The resulting present values from the cash flow projections provided the basis of comparison. This is shown in Table 4–9.

The project present value of the standard Indonesian contract terms (which included the DMO and the IC) fell about halfway between the results of the pure 85%/15% split and the 87%/13% split. The present value of the standard contract terms discounted at 15% is $26.9 million. The pure 85%/15% split had a present value of $30.2 million. The effect of the DMO alone pulled the effective split down to an equivalent 86.4%/13.6% split. This is shown in Figure 4–4.

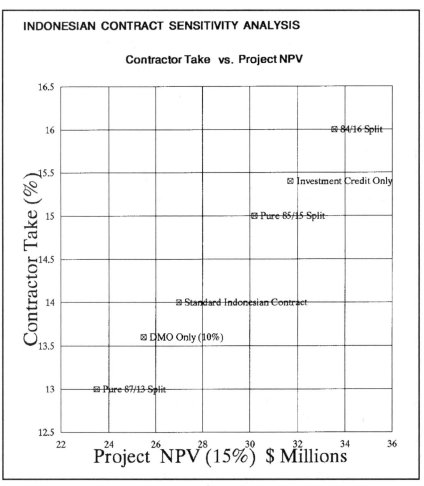

Figure 4-4 Indonesian contract sensitivity analysis

The DMO is nearly invisible from the exploration perspective. Because of the present value effect the DMO also has little influence on development feasibility economics. However, after five years of production, the DMO exerts a stronger influence. Under similar analysis on old oil, the effective split comes closer to an 89%/11% split in favor of the government. This highlights the different perspectives of development feasibility and production economics.

Table 4-8

Indonesian PSC Sensitivity Analysis
Model Input Parameters

INPUT

Field size	50/MMBBLS

Peak production rate	19,000 BOPD
Lambda	14%
(% of reserves produced in peak year)	
Decline rate	12%
Field life	15+ Years

Initial oil price	$18.00/bbl
	4% Escalation

Development capital costs	$108 MM *
	$2.15/bbl

Operating costs	
Fixed	$6 million/year
Variable	$1/bbl
Total	First year $1.85/bbl

RESULTS based upon standard contract terms

Net present value (15% DCF)	$27 million
	$0.52/bbl
Internal rate of return	28%
Return on investment	152%
Payout	5 Years

*50% of capital costs are assumed to be eligible for the 17% investment credit.

Table 4-9

Indonesian PSC Sensitivity Analysis Results

	NPV 15% ($M)	Equivalent Contractor After-tax Take (%)	
Pure 84/16 Split Contractor 30.7692%* No DMO No investment credit	33,543	16.0%	Pure Split
Investment Credit Only Contractor 28.8462%* 17% investment credit No DMO	32,900	15.3%	Effective
Pure 85/15 Split Contractor 28.8462%* No DMO No investment credit	30,192	15.0%	Pure Split
STANDARD CONTRACT Contractor 28.8462%* DMO = 10% of wellhead price 17% investment credit	26,996	14.0%	Effective
DMO Only Contractor 28.8462%* 10% of welhead price No investment credit	25,492	13.6%	Effective
Pure 87/13 Split Nonstandard terms Contractor 25.00%* No DMO No investment credit	23,490	13.0%	Pure Split

Indonesian effective tax rate of 48% assumed in all cases.
* Contractor pre-tax profit oil share

RISK SERVICE CONTRACTS

Service contracts are based on a simple formula: The contractor provides all capital associated with exploration and development of petroleum resources. In return, if exploration efforts are successful, the government allows the contractor to recover those costs through sale of the oil or gas and pays the contractor a fee based on a percentage of the remaining revenues. This fee is often subject to taxes. All production belongs to the government. The net importing countries are the ones most likely to use this approach. They need the crude. By 1994 service agreements were being used in Argentina, Brazil, Chile, Ecuador, Peru, Venezuela, and the Philippines. In Peru either the service contract or a concession could be used.

When the term *service contract* is used it is normally understood to be a *risk* service contract. The term *risk contract* is also used. The term *risk service* is widely accepted but rather inappropriate. The oil service industry would hardly recognize the service contracts found in the upstream end of the business. To refer to an exploration agreement where the oil company puts up all the capital and risks loosing it all as a service agreement is an obvious misnomer. But this is what it is called. The added term *risk* is clearly an improvement. Because the contractor does not get a share of production, such terms as *production sharing* and *profit oil* are not appropriate even though the arithmetic will often carve out a share of revenues in the same fashion that a PSC shares production.

The distinction between PSCs and risk service contracts is minute. The nature of the payment for the contractor's services is the point of distinction. Other than that, the arithmetic and terminology are quite similar. This is why many service agreements are commonly referred to as PSCs. The Philippine risk service contract is a perfect example. It is often referred to as a PSC even by the government. There is, however, a feature found in the Philippine system that makes it unique. The service contracts of the Philippines and Ecuador are examined here.

PHILIPPINE RISK SERVICE CONTRACT

The language of the Philippine contract is identical to that of most PSCs with the exception of the Filipino Participation Incentive Allowance (FPIA). The FPIA is part of the service fee, and it is based on gross revenues just like a royalty except that it goes to the contractor group. The FPIA, based on a sliding scale, can get as high as 7.5% if Filipino participation is 30% or more onshore. Offshore, 15% Filipino participation will qualify for the FPIA.

Filipino Participation (%)	FPIA, %
Up to 15%	0
15–17.5	1.5
17.5–20	2.5
20–22.5	3.5
22.5–25	4.5
25–27.5	5.5
27.5–30	6.5
30 or more	7.5

The Philippine contract has a 70% cost recovery limit, and profit sharing is 60%/40% in favor of the government. The contractor 40% share of profits is not subject to taxation. The contractor's taxes are paid out of the government share of profit oil.

Calculation of the contractor entitlement under the Philippine contract is based on the following assumptions:

> Gross revenues = $100 million
> Assume Contractor Group eligible for full 7.5% FPIA
> Costs eligible for cost recovery =
> $50 million – high cost case
> $20 million – low cost case

CONTRACTOR ENTITLEMENT

Low Cost Case	High Cost Case	
$100.0 MM	$100.0 MM	Gross revenues
–7.5	–7.5	FPIA service fee
92.5	92.5	Net revenues[1]
–20.0	–50.0	Costs recovery
72.5	42.5	Revenues available for sharing
–43.5	–25.5	**Government 60% share**
29.0	17.0	Contractor 40% share
+7.5	+7.5	FPIA
$36.5	$24.5	Total contractor service fee
+20.0	+50.0	Costs recovery
$56.5 MM	$74.5 MM	**Total contractor entitlement**
45.6%	**49.0%**	**Contractor take** [2]

[1]The term *net revenues* is used loosely here.
[2]Total Contractor Service Fee ÷ (Gross Revenues – Cost Recovery)

The government entitlement in the low cost case came to only $43.5 million, or 43.5% of revenues. The contractor's share of gross revenues was $56.5 million. The revenues available for sharing (including the FPIA) came to $80 million ($100 MM – $20 MM cost recovery). Of this, the contractor share came to $36.5 million. The contractor take, therefore, is 45.6% ($36.5 MM/$80 MM).

The contractor take increased from 45.6% to 49% with the higher cost case. There are few systems that behave like this. Because of the way the service fee is calculated, the system is one of the most progressive in the world. This is because the FPIA is based on gross revenues—it behaves like a negative royalty.

ECUADOR RISK SERVICE CONTRACT

Ecuador uses an R factor calculation for its service contract. The contractor's entitlement is based on costs recovery and a service fee that is taxed at a rate of 40%. Part of the fee calculation is based on a formula consisting of a sliding scale R factor. An unusual aspect of the Ecuador service fee is that it is calculated before the normal cost recovery arithmetic found in most PSCs and service agreements. It is not as progressive as the Philippine FPIA, but it is a step in the same direction. The formula for the service fee is as follows:

SERVICE FEE FORMULA:

$$TS = PR(INA) + R(P - C)Q$$

where:

TS = Annual Service Fee payment in U.S. dollars.

PR = Average Prime Rate (decimal fraction)

INA = Development and Production Costs less reimbursements

P = Average International Crude Price ($/bbl)

C = Production Costs ($/bbl)

Q = Annual Production (MMBBLS)
For R Factor Calculation (BOPD)

R = Average Profit Factor (decimal fraction)

= $(R1(Q1) + R2(Q2))/(Q1+Q2)$, etc.

R is based upon the following incremental sliding scale:

Production Rate (Q)	Example R Factor	
Up to – 10,000 BOPD	.30	$R1$ *
10,001 – 30,000 BOPD	.25	$R2$
30,000 – 50,000 BOPD	.23	$R3$
50,000 – 70,000 BOPD	.20	$R4$
70,000 +	.18	$R5$

* The R factor for the first tranche of production ranges from .25–.35 and steps down in increments of .02–.05 depending on the contract.

A sample calculation of contractor entitlement is shown next. It starts with the contractor service fee calculation. The assumptions are outlined as follows:

$$TS = PR(INA) + R(P - C)Q$$

where:

PR = 10% (.10)

INA = $25 million (assumed)

Q = 6 MMBBLS (assumed) = average 16,438 BOPD

R = $(.30 \times 10{,}000 + .25 \times 6{,}438)/16{,}438 = .2804$

P = $16.00/bbl (assumed)

C = $10 MM (assumed) = $1.67/bbl

TS = $.10 \times \$25$ MM + $.2804(\$16.00 - \$1.667) \times 6$MM

= $2.5 MM + $24.114 MM

= $26.61 MM

The contractor entitlement is based upon the after-tax service fee and cost recovery. The step-by-step calculation of the entitlement is shown as follows:

CONTRACTOR ENTITLEMENT

$26.61 MM	Service fee
–2.50	Incentive deduction ($PR \times INA$)
$24.11 MM	Taxable income
-9.64	Income tax (40%)
+2.50	($PR \times INA$)
$16.97 MM	Service fee after-tax
+20.00	Assumed cost recovery
$36.97 MM	Total contractor entitlement

Gross revenues, by the way, were $96 million. The revenues available for sharing came to $76 million ($96 MM – $20 MM cost recovery). Of this, the contractor's share came to $16.97 million, and contractor take therefore is 22.3% ($16.97 MM/$76 MM).

Because of the way the service fee is calculated, the system is fairly progressive. Had the capital costs been higher, the contractor take would also have been higher. This is because of the hierarchy of the calculation and the nontaxable element of the service fee based upon $PR \times INA$. The sliding scale R factor is modestly progressive, but the way the service fee is calculated, the government share flexes upward and downward to accommodate variations in profitability.

RATE OF RETURN CONTRACTS

Contracts with *flexible* terms are becoming standard. There are many advantages for both the host government and the contractor with contracts that encompass a range of economic conditions. This is the acid test.

Table 5-1

Flexible Contract Terms and Conditions

Contract Terms Subject to Sliding Scales	Factors and Conditions That Trigger Sliding Scales
■ Profit Oil Split ■ Royalty ■ Bonuses ▲ Cost Recovery Limits ▲ Tax Rates ◆ Uplifts	■ Production Rates ▲ Water Depth ▲ Cumulative Production ▲ Oil Prices ▲ Age or Depth of Reservoirs ▲ Onshore vs. Offshore ▲ R Factors* ▲ Remote Locations ◆ Oil vs. Gas ◆ Crude Quality (Gravity) ◆ Time Period (History) ◆ Distance from shore ◆ Rate of Return*
■ Most Common ▲ Less Common ◆ Rare	*Rate of Return Contracts

The most common method used for creating a flexible system is with *sliding scale terms*. Most sliding-scale systems trigger on production rates. As production rates increase, government take, in one fashion or another, increases. This theoretically allows equitable terms for development of both large and small fields. Contracts may subject a number of terms to sliding scales.

Some contracts will provide flexibility through a progressive tax rate. Others will tie more than one variable to a sliding scale such as cost recovery, profit oil split, and royalty. Table 5–1 shows the diversity of contract elements that are subject to sliding scales and the factors that trigger a change.

There are many sliding scales. One of the more unusual is the Guatemala contract of the late 1980s with a sliding scale royalty based on variations in crude quality. The royalty rate increases or

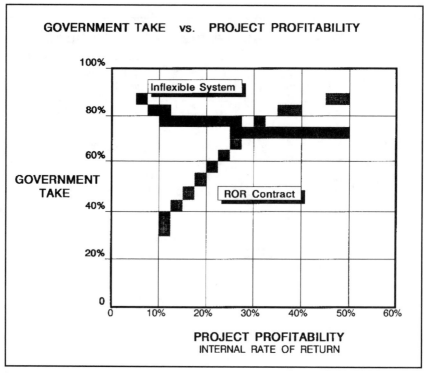

Figure 5–1 Government take vs project profitability

decreases one percentage point every degree above or below 30° API crude gravity.

The objective with sliding-scale systems is to create an environment where the government take flexes upward with increased profitability. The result of most fiscal structures though is that project profitability is a function of government take. As a rule, it is better for both parties when government take is a function of profitability. Figure 5–1 illustrates this aspect of flexibility, showing how government take increases as project profitability increases with a flexible rate of return (ROR) system. This is the objective of sliding scales as well as ROR systems. Inflexible systems with high royalty rates can work in just the opposite way. A stiff nonnegotiable royalty is the antithesis of flexibility. Even a progressive sliding-scale royalty scheme can be regressive. Just

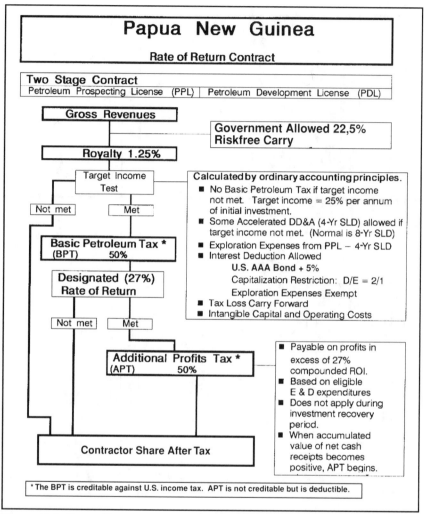

Figure 5–2 Papua New Guinea regime

because the royalty rate becomes progressively larger with some pseudomeasure of profitability, such as production rates, royalties are so strongly regressive that with most marginal discoveries, government take is invariably higher. There are many systems that exhibit this characteristic of *reverse flexibility*.

Some countries have developed progressive taxes or sharing arrangements based on project rate of return (ROR). The effective

Table 5-2

Sample Rate of Return Contract Cash Flow Projection

Year	OIL Production (MBBLS) (A)	OIL Price ($/bbl) (B)	Gross Revenues ($M) (C)	Tangible Cap Ex ($M) (D)	Operating Expense ($M) (E)	DD&A ($M) (F)	Deductions ($M) (G)	Taxable Income ($M) (H)	Basic Income Tax ($M) (I)	Net Cash Receipts ($M) (J)	Amount Brought Forward ($M) (K)	Amount Carried Forward ($M) (L)	Additional Profits ($M) (M)	Resource Rent Tax ($M) (N)	Net Cash Flow ($M) (O)
1994	0	$20.00	0	20,000	0	0	0	0	0	(20,000)		(20,000)	0	0	(20,000)
1995	0	20.00	0	8,000	0	0	0	0	0	(8,000)	(26,000)	(34,000)	0	0	(8,000)
1996	0	20.00	0	50,000	0	0	0	0	0	(50,000)	(44,200)	(94,200)	0	0	(50,000)
1997	4,500	20.00	90,000	25,000	11,500	20,600	32,100	57,900	20,265	33,235	(122,460)	(89,225)	0	0	33,235
1998	7,000	20.00	140,000	0	14,000	20,600	34,600	105,400	36,890	89,110	(115,992)	(26,882)	0	0	89,110
1999	5,600	20.00	112,000	0	12,600	20,600	33,200	78,800	27,580	71,820	(34,947)	0	36,873	18,436	53,384
2000	4,760	20.00	95,200	0	11,760	20,600	32,360	62,840	21,994	61,446	0	0	61,446	30,723	30,723
2001	4,046	20.00	80,920	0	11,046	20,600	31,646	49,274	17,246	52,628	0	0	52,628	26,314	26,314
2002	3,439	20.00	68,782	0	10,439	0	10,439	58,343	20,420	37,923	0	0	37,923	18,961	18,961
2003	2,923	20.00	58,465	0	9,923	0	9,923	48,541	16,990	31,552	0	0	31,552	15,776	15,776
2004	2,485	20.00	49,695	0	9,485	0	9,485	40,210	14,074	26,137	0	0	26,137	13,068	13,068
2005	2,087	20.00	41,744	0	9,087	0	9,087	32,657	11,430	21,227	0	0	21,227	10,613	10,613
2006	1,732	20.00	34,647	0	8,732	0	8,732	25,915	9,070	16,845	0	0	16,845	8,422	8,422
2007	1,427	20.00	28,411	0	8,421	0	8,421	19,990	6,997	12,994	0	0	12,994	6,497	6,497
2008	0	20.00	0	0	0	0	0	0	0	0	0	0	0	0	0
	40,000		799,864	103,000	116,993	103,000	219,993	579,870	202,955	376,916			297,624	148,812	228,104

(A) Production Profile
(B) Crude Price
(C) Gross Revenues = (A) x (B) (No royalty in this example)
(D) Tangible Capital Costs [Capitalized — see Column (F)]
(E) Operating Expenses (Expensed)
(F) Depreciation of Tangible Capital Costs: 5-Year Straight Line Decline
(G) Deductions = (E) + (F): Up to maximum of Net Revenue
    (However, this example assumes no royalty)
(H) Taxable Income for Basic Income Tax = (C) − (G)

(I) Basic Income Tax = 35%
(J) Net Cash Receipts = (C) − (D) − (E) − (I)
(K) Amount Brought Forward = (L from previous year) x 1.3 [Amount Carried Forward
    uplifted by 30%—This is the ROR aspect]
(L) Amount Carried Forward = (J) + (K) if the balance is still negative
(M) Additional Profits = (J) + (K) if the balance is positive
(N) Resource Rent Tax = (M) x .5
(O) Contractor After-tax Net Cash Flow = (C) − (D) − (E) − (I) − (N)

government take increases as the project ROR increases. In order to be truly progressive, the sliding-scale taxes and other attempts at flexibility should be based on profitability, not production rates. Most contracts have progressive elements, but they are usually based on levels of production instead of a direct measure of profitability. Production levels are a proxy for profitability, but that is all. There are many other factors that influence project profitability, which is why ROR contracts are structured the way they are. ROR systems take into account product prices, costs, timing, and production rates. All of these factors influence project profitability.

The ROR approach is characterized by a modest royalty and tax. The state receives no other funds until the oil company has recovered the initial financial investment plus a predetermined threshold rate of return. Theoretically, this rate of return would represent a minimum rate to encourage investment. The government share is calculated by accumulating the negative net cash flows and compounding them at the threshold rate until the cumulative value becomes positive. When that happens, additional taxes are levied, but the contractor still receives some of the profits in excess of the threshold rate of return. These additional taxes are often referred to as resource rent taxes (RRT).

THE PAPUA NEW GUINEA ROR SYSTEM

The PNG system is typical of the classic ROR formula. Under this system the government receives a 1.25% royalty and a 22.5% carried interest (carried through exploration). A basic petroleum tax (BPT) of 50% is levied only if the contractor's income meets or exceeds 25% of the initial investment.

There is an additional tax levied if the contractor's rate of return exceeds 27%. This is done by compounding the negative net cash flows at a rate of 27%. Once the cumulative net uplifted cash flow becomes positive the additional 50% resource rent tax kicks in. It is called the Additional Profits Tax (APT). This is the hallmark of a ROR system. It is also called a *trigger tax*. Reaching a minimum rate of return triggers the tax. The basic structure of the PNG contract is illustrated in Figure 5–2.

Calculation of cash flow from a simple ROR contract is shown in Table 5–2. It outlines the basic ROR system elements with a detailed explanation of the calculations involved in arriving at year-by-year cash flow. In this example, there is no royalty, and the basic income tax is 35%. A 30% uplift is applied on the accumulated negative net cash flows. Once the cumulative balance of net cash flow becomes positive, an additional 50% resource rent tax is imposed.

Critics of the ROR concept complain that these contracts are too restrictive, that the uplift (rate-of-return) places an unreasonably low ceiling on upside potential.

The resource rent tax concept was first employed in Papua New Guinea (PNG). Other countries that use this kind of tax are Australia, Liberia, Equatorial Guinea, and Tanzania.

The treatment of interest cost recovery is close to the ROR concept. Interest cost recovery features found in some PSCs or concessionary systems normally apply a compound interest rate to unrecovered capital costs. In many ways this builds in an element of guaranteed return on invested capital. The similarity ends there because there is not an additional tax once the uplifted cost pool has been recovered.

TUNISIAN CONTRACT WITH *R* FACTOR

Some contracts use what is called an *R* factor. The most common use of such a factor is found in the Tunisian and Peruvian contracts. In these contracts the definitions are virtually identical: *R* factor = Accrued Net Earnings/Accrued Total Expenditures. In Tunisia oil and gas royalties, taxes, and government participation are all based upon the *R* factor.

TUNISIAN *R* FACTOR

$$R = \frac{X}{Y}$$

where:

X = Cumulative net revenue actually received by the contractor equals *turnover* (gross revenues) for all tax years less taxes paid

Y = Total cumulative expenditure, exploration, and appraisal expenses and operating costs actually incurred by contractor from date contract is signed

Example: **Tunisian *R* Factor/Sliding Scale Taxation**

R Factor	Income Tax Rate, %
< 1.5	50%
1.5–2.0	55
2.0–2.5	60
2.5–3.0	65
3.0–3.5	70
3.5 +	75

In this contract the *R* factor is based on a return on investment (ROI). Once the contractor has received his costs plus 50%, or a ROI of 150%, the tax rate increases from 50% to 55%. In some respects it is similar to a rate-of-return contract. The typical rate of return contracts trigger on internal rate of return (IRR). These concepts are discussed later.

Colombian *R* Factor

In 1994 the Colombian government introduced yet another variation on the *R* factor theme. The formula started out relatively similar to other such factors with one twist. The definition of the Colombian *R* is as follows:

Colombian *R* Factor

$$R = \frac{X}{(ID + A - B + (\alpha \times C) + GO)}$$

where:

X = Accumulated earnings of the contractor

$(ID + A - B + (\alpha \times C) + GO)$ = Accumulated Investment + Accumulated Costs of the contractor

ID = 50% of cumulative gross development costs

A = Cumulative gross successful exploration costs

B = Cumulative successful exploration costs reimbursed by ECOPETROL (50% partner)

C = Cumulative gross unsuccessful exploration costs

α = Proportion of dry-hole costs reimbursed by ECOPETROL—a bid item, maximum 50%

GO = Contractor cumulative net operating costs including War Tax payments and duties on imports

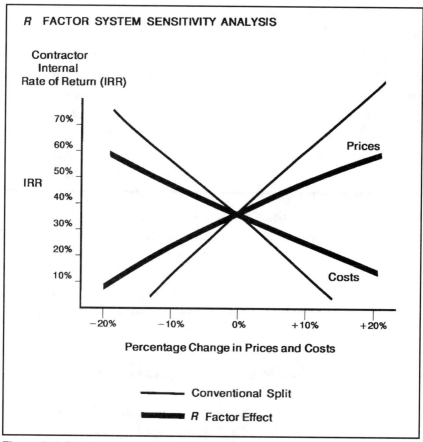

Figure 5–3 *R* factor system sensitivity analysis

Colombian Sliding Scale *R* Factor

R Factor	Contractor Percentage Participation
< 1.0	50%
1.0–2.0	50/*R*
2.0 +	25

The contractor share is 50% after royalties until payout is achieved. If *R* = 1.5, the contractor share equals 33.33% and continues to step down until it reaches 25%, when two times payout is achieved.

Figure 5–3 illustrates the effect an R factor can have on project economics. The results on contractor project IRR are shown as costs and oil prices vary. Costs and prices are the most sensitive factors in project economics. They have the largest impact, and with an R factor, both are accounted for simultaneously. The R factor deals with all variables that affect project economics. The sensitivity is shown by the slope of the lines. The R factor has a dampening effect. Contractor potential upside from price increases is diminished, but the downside is also protected. It is the same with costs. If costs are relatively higher, the R factor mitigates the negative impact. If costs are lower both the contractor and government benefit.

Some governments have devised unique and complex fiscal elements as seen in some of these sample risk service agreements and Illustration 5–1.

JOINT VENTURES

Joint ventures are common in the international oil industry. Most companies are willing to take on partners for large-scale or high-risk ventures in order to diversify—this is good risk management. These joint operations between industry partners differ from the government-contractor relationships that are also joint ventures but are normally referred to as *government participation*.

Some of the proposed Russian joint ventures are characterized by a 100% carry for the production association partner through development including operating costs. This is an extreme example of government participation. However, most of the Russian JVs deal with proved, well-delineated reservoirs. The exploration risk aspect is greatly diminished.

The opening up of the former Soviet Union and other countries dominated by centrally planned economies has added dimension to the joint-venture concept. These countries, particularly republics of the former Soviet Union and Eastern Europe, prefer joint ventures

Illustration 5–1 fiscal creativity

because they have personnel and organizations in place that need to be integrated into future operations. Many areas with petroleum potential in these countries already have petroleum operations with equipment, infrastructure, and personnel.

Figure 5–4 illustrates the general nature of a contractor/government joint venture. Here the government (through the national oil company) is a 30% working-interest partner. The proceeds in this example are subject to the terms of a PSC with a 60%/40% profit oil split in favor of the contractor group. However, the contractor group includes the government as partner. Both partners receive their prorated share of cost oil. Profit oil is split according to the working interest shares. This example shows the profit oil split according to the PSC and the additional split dictated by the joint venture arrangement. This is why many people treat government participation as though it were an added layer of taxation. However, when the government backs in after discovery, it is

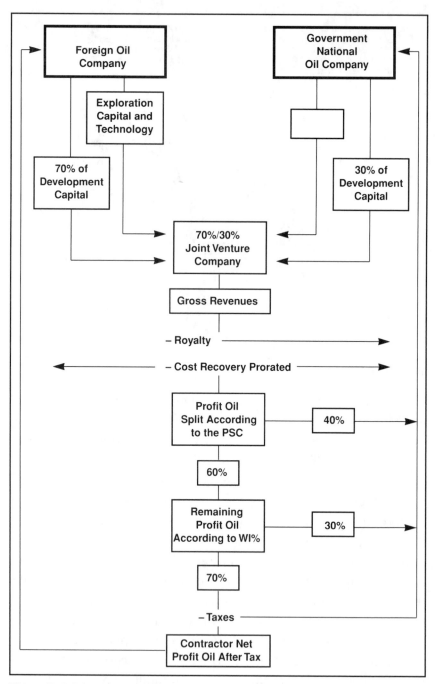

Figure 5–4 Joint venture, PSC

effectively cashing in on the value added at that point. That element of government take occurs long before production begins.

In a pure joint venture, the host government and the contractor share equally in costs and risks. This would have little practical application. However, there is a broad range from pure joint ventures to some of the least pure joint enterprises found in the former Soviet Union. The key ingredient is the amount of carry.

Under most oil company/government joint ventures the oil company bears the costs and risks of exploration. In other words, the government is carried through exploration. This is fairly normal and is automatically assumed whenever some percentage of government participation is quoted.

JOINT VENTURE/GOVERNMENT PARTICIPATION SPECTRUM

Light

— **Pure Joint Venture**
All costs/risks shared
Very rare

— **Mauritania Type Participation**
Government carried through exploration
Contractor recovers exploration costs
plus 50% uplift on government share

— **Typical Joint Venture**
Government carried through exploration
Most common
Contractor can recover exploration costs

BURDEN

— **Colombian Type Joint Venture**
Government carried through exploration
and delineation

— **Full Carry**
Government carried through exploration
and development
Not common

Heavy

— **Russian Type of Joint Venture**
Government carried through rehabilitation
and development, until it has cash flow
from operations

105

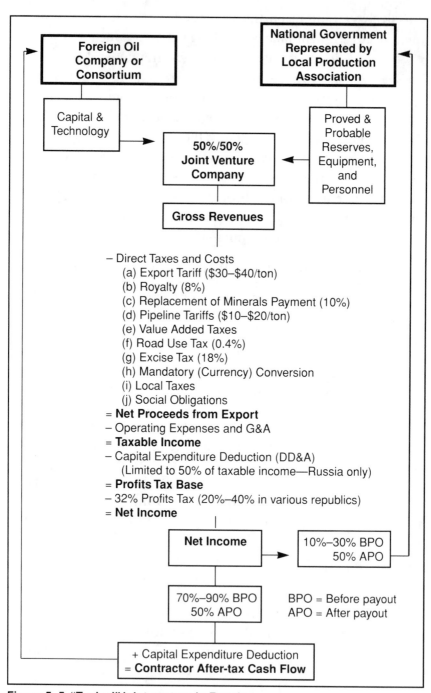

Figure 5–5 "Typical" joint venture in Russia

Where the government actually pays its share of costs, the government share of profits is not the same as a tax on income. When government is carried through exploration and backs in after discovery, the impact of government participation behaves like a capital gains tax. In extreme cases (like Russia) where the contractor pays all rehabilitation, development, and operating costs, the government share of joint-venture profits behaves like an added layer of taxation. At any rate, all forms of government participation reduce the potential rewards of exploration. This must be factored into the risk equation.

Direct reimbursement vs. cost recovery is a present value issue. The contractor invariably recovers costs of exploration and development either through cost recovery, deductions, or direct reimbursement. The difference is timing.

Figure 5–5 outlines a sample Russian joint venture, if there is such a thing. With the proliferation of negotiated deals, it is extremely hard to characterize them all with one example. However, this particular structure has many of the basic elements. A 50%/50% joint venture has strong appeal to the Russians. Figure 5–5 outlines a fairly harsh fiscal system. It is probably too harsh to survive, but this is the often-proposed basis for initial negotiations. Many elements are based on gross revenues and act like royalties regardless of what they are called. Net income is divided between JV partners 90%/10% to 70%/30% in favor of the foreign oil company before payout (BPO) and 50%/50% after payout (APO). Often profits are reinvested BPO. The government share is viewed by many as just another layer of taxation, and a large one at that. The contractor take ranges from 10%–20%.

TECHNICAL ASSISTANCE CONTRACTS (TACs)

TACs are often referred to as rehabilitation, redevelopment, or enhanced oil recovery projects. They are associated with existing fields of production and sometimes, but to a lesser extent, abandoned fields. The contractor takes over operations including equipment and personnel if applicable. The assistance that includes capital provided by the contractor is principally based on special technical know-how such as steam or water flood expertise.

Many smaller companies are finding their niche by targeting these projects. They provide lower-risk situations with opportunities for a company to leverage technical expertise. Despite the reduced risks, EOR projects require careful screening. Some countries have fields and basins that are nearly as depleted as those found in the United States. Few rehabilitation/EOR projects beyond the primary recovery stage will be viable. Added potential such as infill drilling or undrilled deeper horizons often provides the primary financial incentive. Cost and timing estimates are critical, and fiscal terms are critical.

Indonesia has special terms for *marginal* fields (those with production of less than 10,000 BOPD of production). A 10,000-BOPD field could have 15–25 MMBBLS of oil.

Many countries try to tighten the fiscal terms on EOR projects because of the reduced risk. This can be seen in the various contracts offered in Myanmar (Appendix A), where the proposed profit oil split for exploration blocks is 65%/35% in favor of the government, while the proposed split is 70%/30% for improved oil recovery (IOR/EOR) blocks. Unfortunately, most companies do not agree that EOR projects can sustain tougher fiscal terms.

If fiscal terms are out of balance, no amount of technical expertise will salvage a project. As a rule, it takes a minimum of 10–30 MMBBLS of recoverable reserves to justify the costs of an international development project—usually more for EOR. And, if the project is an EOR project that requires substantial pressure makeup, then the fiscal terms need to be flexible.

Development threshold field size for EOR is generally about twice what it would be for primary production. An EOR project in Western Siberia may well require 150 MMBBLS or more of recoverable reserves to create sufficient economy of scale.

If there is existing production, then a decline rate, or production profile, is negotiated. The production defined by the negotiated decline rate is exempt from the sharing arrangement. This production goes directly to the government. Production above the negotiated rate is *incremental* production, which theoretically is the result of the technical assistance provided by the contractor. The concept is illustrated in Figure 5–6. This incremental production is normally subject to a production sharing arrangement, although TACs can be found under a variety of systems.

Key elements of these arrangements:
• Government need for technology and capital
• Associated personnel
• Existing proved reserves
• Existing infrastructure, and equipment
• Joint management

Rehabilitation projects are often structured in three phases that follow a logical sequence. Each phase carries a specific work program, and the contractor has the option to proceed into each subsequent phase. The decision to go forward is based on the technical results of each previous phase. A three-phase TAC is outlined as follows:

THREE-PHASE TECHNICAL ASSISTANCE CONTRACT OUTLINE

PHASE ONE: FEASIBILITY STUDY
• Bonus
• Minimum work commitment
• 6 months to 1 year

At end of the feasibility study, the contractor has the option of surrendering the acreage and dropping out if the work commitment has been fulfilled. Or the contractor may choose to go forward into Phase Two. The government is given the results of the feasibility study and presented with a work program outline and a budget for the next phase.

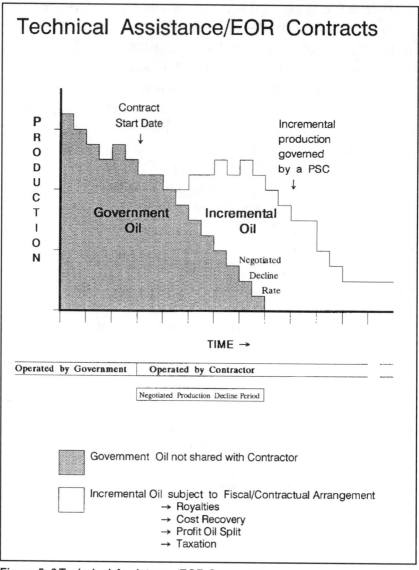

Figure 5–6 Technical Assistance/EOR Contracts

PHASE TWO: PILOT PROGRAM

- Bonus
- Minimum work program
- 2–3 years

The pilot program will consist of a small-scale water or steam injection test to determine if the reservoir will take injected fluids at sufficient rates. During the injection the contractor will monitor the reservoir response. At the conclusion of the pilot program, if the work program is fulfilled, the contractor has the option of relinquishing all rights or entering Phase Three, the commercial development phase. The government is given the results of the second phase pilot program and presented with budget for the next phase.

PHASE THREE: COMMERCIAL DEVELOPMENT
- Bonus
- Workovers
- Drilling
- Implement full-scale EOR

If the contractor enters into the commercial development stage of the contract, production will be shared through a PSC governed by a royalty/tax system or divided according to a specific joint-venture arrangement.

HALLWOOD CASPIAN

In April 1994, Hallwood Energy Partners announced by news release it had entered an EOR joint-venture agreement with the State Oil Company of the Azerbaijan Republic (SOCAR) for the Mashal Field located in the Caspian Sea. Discovered in 1954, the field had already produced over 50 MMBBLS of oil. Production at the time the contract was signed was 1,750 BOPD. The terms of the deal disclosed by Hallwood indicated that Hallwood and SOCAR would recover their costs proportionally from 70% of net revenues from production above 2,700 BOPD. During the first phase of the joint venture, profit would be divided 60% to Hallwood and 40% to SOCAR. The joint-venture term was for 25 years with an initial term of three years, which called for Hallwood to spend a maximum of $65 million and a minimum of

$5 million in a three-phase development plan for the field. There are aspects found here that are characteristic of some of the rehabilitation deals found in the former Soviet Union.

Cost Recovery

In many proposed joint ventures in the FSU republics, the government partner also wants to recover costs. From the perspective of the industry partner, the government cost recovery can act like an added layer of taxation. If SOCAR is also putting up money, then it would not be viewed that way. However, it is unlikely that SOCAR has any money to provide for the project. The amount of government cost recovery for past costs is a point of debate/negotiation. The government cost recovery would likely be funded out of incremental production.

Cost recovery is also limited to 70% of net revenues, which unsurprisingly implies a royalty. Hallwood, therefore, would be able to recover costs out of something less than 70% of gross revenues.

Incremental Production

The incremental production starts at 2,700 BOPD, not at the current rate of production that was announced as 1,750. Therefore, 950 BOPD of production must be made up before cost recovery or profit sharing may begin. Many negotiations for these deals can get sticky over a beginning like this. The extra 950 BOPD would be viewed by some as an additional royalty.

Net Revenues

The release indicated that sharing would be 60%/40% during the first phase of the joint venture. This implies that after payout the sharing arrangement will change in favor of the government. Furthermore, the Hallwood share of profit will likely be subject to taxes.

There is not sufficient information to determine if it is a good deal or not. The exploration rights may be of value. Perhaps there is life left in the field.

6

THRESHOLD FIELD SIZE ANALYSIS

Threshold field size analysis is an important exercise in the international business. It focuses exploration efforts and provides insight into fiscal systems. It also helps define boundary conditions for explorers and development engineers. Too often, time is spent evaluating the prospects in a country that is dominated by a fiscal system that simply will not justify exploration efforts.

The subject of success probability centers on the difference between technical success and commercial success. The difference, as mentioned earlier in this book, is development threshold field size. As the threshold field size approaches zero, as it has in the United States for all practical purposes, the difference begins to disappear. It has not disappeared in the international sector. This is one reason why flexible contracts have been created.

Development thresholds depend on costs, lead times, reservoir characteristics, and the host country fiscal system. The main reason for threshold analysis is to decide whether or not to even attempt exploration efforts.

Risk capital is a critical factor, and it depends on the nature of the work commitment. Companies must keep exploration costs down in order to compete. But it is hardly an issue in development economics.

EXPLORATION VS. DEVELOPMENT THRESHOLDS

The huge risks force oil companies to search for giant fields. It is a basic formula for the industry. But once a discovery has been made, the amount of risk associated with finding it has little meaning. The decision to develop a discovery is determined by the technical and economic feasibility from that point forward.

Development geologists and engineers can develop many fields that would not have been considered large enough to justify risking exploration capital. Threshold field size, from an explorationist's point of view, is on the order of 100–300 MMBBLS or more, but development thresholds are an order-of-magnitude smaller.

The key variables used in exploration threshold field size analysis are:

- Risk capital estimate
  (Work commitment, G&G, seismic, dry-hole costs, etc.)
- Probability of finding hydrocarbons
- Estimated development costs
- Oil prices estimates
- Expected field size distribution
- Estimated production profiles
- Fiscal terms

The key variables used in development threshold field size analysis are:

- Estimated development costs
- Oil prices estimates
- Estimated reserves and production profiles
- Fiscal terms
- Sunk costs

Oil prices and costs are some of the most sensitive factors. Per-well deliverability is also one of the key factors because it has such a strong influence on capital and operating cost require-ments. Costs are also strongly affected by water depth and remoteness.

Discounted cash flow analysis is the workhorse of threshold field size analysis. The deepwater and remote frontier regions pro-vide good examples for demonstrating methods of determining threshold field size. From a geographical perspective, 600 ft of water, as a rule, marks the limit of the continental shelf and the beginning of the continental slope. Indonesia and Malaysia offer special deepwater incentives for projects in water deeper than 600 and 650 ft, respectively. In the international sector, even 400 ft of water can seem awfully deep.

This industry is characterized by high capital costs and diverse technical boundaries. Yet technical feasibility is becoming less of an issue. Field-proven off-the-shelf technology provides a wide array of building blocks for frontier field development under a variety of conditions. The most important challenge at the moment is to reduce the high capital and operating costs associat-ed with frontier regions.

In many regions offshore seismic costs for data acquisition, processing, and interpretation can range from $1,000 to $1,500/km. In the remote jungles of Southeast Asia, the acquisi-tion costs alone can exceed $20,000/km, and in the highlands of Papua New Guinea, the costs are even higher. Imagine 100 km of seismic data costing over $2 million!

Drilling costs depend on many factors, but they are magnified in the international business. Many wells in the Timor Gap were budgeted at around $9 million each for the relatively moderate water depths. In the Philippines many of the deepwater wells cost over $25 million.

Mobilizing a rig and services into a remote onshore location can easily put well costs for a modest depth well with no harsh drilling conditions into the $5 million range. Building a drilling

115

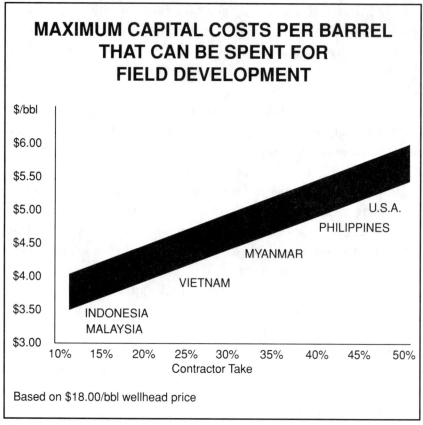

Figure 6–1 Comparing fiscal terms: Capital cost limits

location and a road to the wellsite alone can rival the costs of some U.S. wells. The Llanos of eastern Colombia is characterized by deep, tough drilling conditions. A 17,000-ft well can easily cost $20 million and stands a strong chance of not reaching its objective due to steeply dipping beds and hard, deep drilling.

OPERATING COSTS
VS.
CAPITAL COSTS

The tradeoff between operating costs and capital costs is dominated by the timing of these costs. Operating costs are spread over such a much longer time period and follow most of the capital cost expenditure. Therefore, in terms of present value, capital costs are twice as critical as operating costs.

Some analysts use a *technical cost factor* that combines capital and operating costs in terms of dollars per barrel. The formula for the technical cost factor is:

Technical cost factor = Unit capital cost
+ Unit operating cost ÷ 2

For example, if $100 million is required to develop a 50-MMBBL field, the capital cost is $2.00/bbl. If during the expected 15-year life of the field, projected operating costs average $10 million per year, the unit operating cost then would be $3.00/bbl [(15 years × $10 MM/year)/50 MMBBLS]. The technical cost index then would be $3.50/bbl [$2.00 + ($3.00/2)].

Figure 6–1 shows how the contractor take correlates with capital cost limits for field developments. For example, in Indonesia, as capital costs approach $3.50/bbl project economics quickly become subeconomic at an $18.00/bbl wellhead price.

TECHNICAL VS. COMMERCIAL SUCCESS RATIOS

Consider an area where technical success ratios are on the order of 20% or more, like much of the Gulf of Mexico and many areas in Southeast Asia. Success ratios are enhanced these days with high-quality seismic data, especially offshore, but the difference between technical success and commercial success can be substantial. This is particularly the case in deepwater and frontier regions. This is illustrated in Figure 6–2. In this example, a gas discovery is not considered viable, and unfortunately this is often the case. Gas development thresholds are discussed later.

For many areas worldwide, the frontier environments will require a bare minimum of 50 MMBBLS of recoverable oil to justify development. To consider anything less than a 50-MMBBL field as noncommercial can have a strong influence on the commercial success ratio. If the probability of finding hydrocarbons is on the order of 20%, then it would not be unrealistic to expect the probability of finding something greater than 50 MMBBLS to drop down to 10% or less. In the Gulf of Mexico deepwater, the probability of finding a field larger than 50 MMBBLS of oil or oil equivalent (BOE) is 5%–8%. Commercial success ratios are further limited in frontier areas, because gas discoveries must be huge to be economic.

From the viewpoint of the explorationist, the question is, "How large must the target be?" High deepwater and frontier costs require huge targets. Furthermore, tight fiscal terms demand even larger targets to justify spending risk dollars.

The Expected Monetary Value (EMV) approach introduced in Chapter 2 is used in a slightly different way to quantify the explorer's position. The relationship between risk dollars and the anticipated probability of success is outlined with the EMV equation.

Threshold field size (reward) is estimated by making assumptions about the probability of success from exploration drilling and the value of the reserves that may result from successful exploration efforts. Assuming that discovered undeveloped reserves are worth around 50¢/bbl, and exploration costs are U.S. $15 million, the relationship between an assumed success ratio of 8% and required target size can be calculated. The EMV at breakeven is zero. This identifies the threshold or minimum target size for exploration efforts under the assumption outlined above.

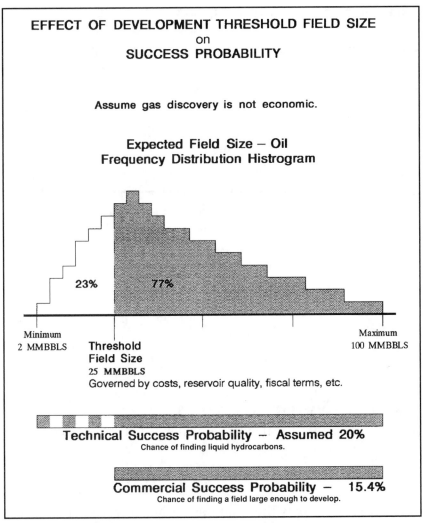

EFFECT OF DEVELOPMENT THRESHOLD FIELD SIZE
on
SUCCESS PROBABILITY

Assume gas discovery is not economic.

Expected Field Size — Oil
Frequency Distribution Histrogram

23%   77%

Minimum
2 MMBBLS   **Threshold**   Maximum
              Field Size   100 MMBBLS
              25 **MMBBLS**
Governed by costs, reservoir quality, fiscal terms, etc.

Technical Success Probability — Assumed 20%
Chance of finding liquid hydrocarbons.

Commercial Success Probability — 15.4%
Chance of finding a field large enough to develop.

Figure 6–2 Technical vs. Commercial success probability

EXPECTED MONETARY VALUE BREAKEVEN ANALYSIS

EMV = (**Reward** $\times SP$) – [**Risk capital** $\times (1 - SP)$]

where:

EMV	=	Expected Monetary Value
	=	0 at breakeven
Risk capital	=	$15 million
SP	=	Probability of success = 8%
Reward	=	Minimum target size (in this case)

0 = (Reward $\times$ .08) - [$15 MM $\times$ (1 - .08)]

(Reward $\times$ .08) = [$15 MM $\times$ (1 - .08)]

$$\text{Reward} = \frac{\$15\ \text{MM} \times .92}{.08}$$

Reward	=	$172.5 million
	=	Breakeven results
	=	Threshold value
	=	345 MMBBLS (at 50¢/bbl)

Figure 6–3 shows the relationship between risk dollars and success ratio. If $15 million are placed at risk with only 8% chance of success, then the potential reward must be worth at least $172.5 million. Figure 6–4 describes what this means in terms of exploration threshold field size for deepwater in Southeast Asia and the Gulf of Mexico. These target sizes are consistent with a technical success ratio of perhaps up to 20% and development thresholds of over 50 MMBBLS.

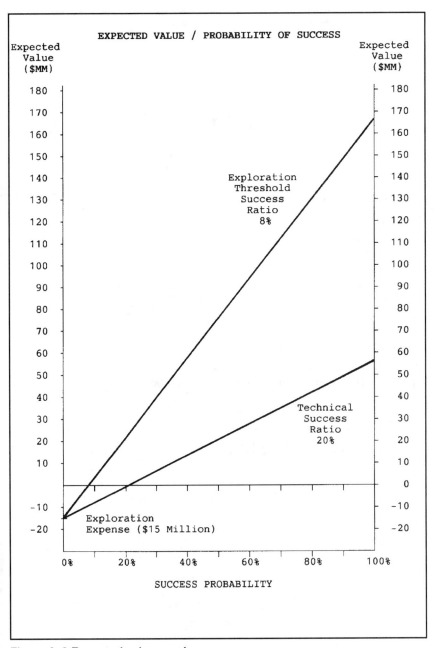

Figure 6–3 Expected value graph

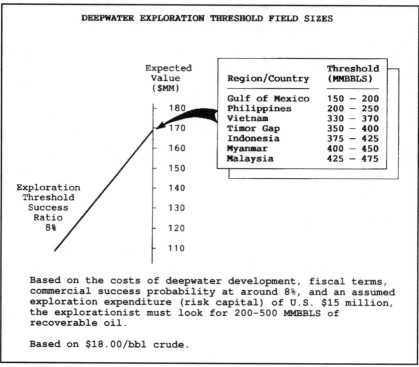

Figure 6–4 Detail: Breakeven/Threshold exploration target

Table 6–1

Wave Height Comparison, feet

	100-year Wave	50-year Wave	Average Winter	Summer
North Sea	90	85	8	5.5
Gulf of Mexico	60	50	5	4
Java Sea	30	25	2.6	2.3

GULF OF MEXICO VS. SOUTHEAST ASIA COMPARISON

Much of the international business is offshore. Over 50% of the activity in Southeast Asia (excluding China) in terms of production and drilling is offshore. The Gulf of Mexico is a good

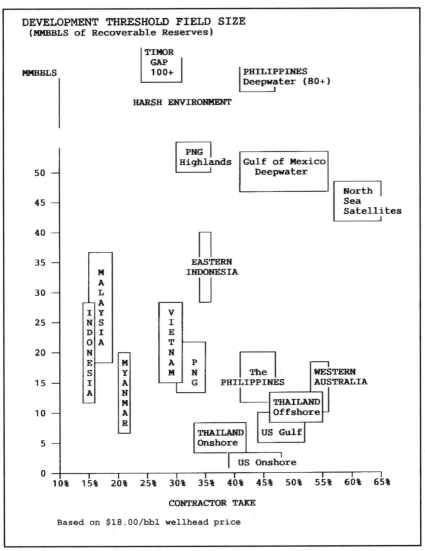

Figure 6–5 Minimum Recoverable Reserves for Development

source of comparison for much of that region. Generally, the climates are similar. By contrast, the North Sea is quite harsh in terms of temperature, water depth, and wave height. Table 6–1 compares wave height characteristics.

The kind of exploration thresholds found in the region are

illustrated in Figure 6–4. Figure 6–5 shows development thresholds. The systems that have little or no flexibility have limited expression along the X axis. Some systems are quite flexible, and contractor take can have quite a broad range. Other systems are fairly rigid—like Malaysia and Indonesia. Indonesia developed a new set of terms for the Eastern frontier in 1994. In that region the split is 65%/35% in favor of the government.

Reservoir quality and per-well deliverability are nearly as important as total reserves. Project economics in the deepwater and frontier regions require prolific production rates.

Special deepwater and frontier terms are critical because of the risks and costs of deepwater exploration and development. The incentives currently in place in some countries are not sufficient.

GAS PROJECTS

In international exploration, oil and gas are quite different. Gas discoveries are often noncommercial unless they are quite rich in liquids, close to an existing market, or very large. There are many giant gas fields that are still *waiting on pipe*.

Threshold field size for gas is substantially greater on a Btu basis for both exploration and development thresholds. For example, in some areas the threshold field size for development of an oil discovery may be on the order of 20–30 MMBBLS. This situation is shown in Figure 6–5. In the same region, gas development threshold field sizes may be more on the order of 500 BCF to over 3 TCF (100–500 MMBOE).

Contract terms for oil are clearly defined, yet the terms for a possible gas discovery may be quite vague. In many systems where gas terms are nailed down, they probably should not be. Sometimes a simple *gas clause* is used, indicating that if gas is discovered, the government and the contractor will sit down and negotiate.

In many places there are substantial quantities of associated gas that are simply flared every day. In the famous Russian

Samotlov oil field, gas was flared at rates of up to 2 BCFD. At the AMOCO-operated gas/condensate Sajaa field and liquids extraction plant in Sharjah, up to 450 MCFD were flared at one time.

When gas discoveries do get developed, it is normally quite a long time after discovery. In the ARCO-operated Northwest Java (NWJ) block off the coast of Java, a gas development project exceeding $1 billion finally was launched in 1993 for associated and nonassociated gas. Prior to that, for over 20 years, the nonassociated gas was simply flared—up to 100 MMCFD at times. The same amount of gas was being flared in the adjacent Southeast Sumatra (SES) block.

If the gas had not been flared, the oil could not have been produced economically. There are numerous countries that are tightening their flaring policy, and the flaring option is becoming rare. But shut-in gas is nearly worthless and so is shut-in oil.

When a gas discovery is made, it is customarily followed by a well-known ritual. What can we do with all this gas? Most explorationists are oil oriented. Gas is not a stranger, but gas developments in the international arena are generally large scale, and the numbers get big quickly. What are the options? The list includes:

- Gas Sales: produce it and pipe it to market
- LPG: liquids extraction, gas plant
- Methanol
- Fertilizer: ammonia/urea
- LNG

GAS SALES

The ideal situation for a gas discovery would be a local market for the gas at a reasonable price. The United States and Europe are about the only places where these things are taken for granted. Most exploration acreage is located a long distance from the kind of markets that would make gas sales a simple matter of laying pipe. Numerous gas discoveries have sat idle for years awaiting development.

By as early as 1983, ARCO knew it had at least 3 TCF of recoverable gas in the Yacheng discovery just south of Hainan Island, and it could at that time guarantee long-term deliverability of at least 400 MMCFD for over 20 years. The purchase agreement and development plans waited 10 years. Building a 10-year wait into exploration economics can kill a lot of drilling decisions. And, it will have been much longer than 10 years from the spudding of the discovery well to first deliveries in Kowloon 480 mi and U.S. $1.2 billion away. First deliveries were scheduled for January 1996, 13 years after discovery.

LIQUIDS EXTRACTION AND LPG

Liquids extraction can range from low-volume plants that strip out condensates to large-scale facilities that liquify LPGs. LPGs are primarily de-ethanized propanes and butanes with some pentanes thrown in. Condensates are made up of pentanes and some of the heavier hydrocarbons (see Appendix I). Worldwide gas plant sizes range from 98 to 124 MMCFD in terms of inlet capacity, and liquid/gas yield averages 26 bbl per MMCF of condensate.

The international sector requires rather large-scale projects, and if gas is rich enough, then LPG extraction may be a development option. As a rule, the condensate yield alone must be at least 30 bbl per MMCF, and LPGs may double that. LPG fractionating plants can be big projects.

An LPG facility at the ARUN LNG complex in Indonesia cost $400 million to construct in the late 1980s. The capacity was 1.9 million tons/year—over 55,000 B/D. In late 1993 the Gas Authority of India announced plans to build a 140,000 metric ton/year LPG plant (6,000–7,000 B/D) in Maharashtra (*Oil & Gas Journal*, Nov. 8, 1993, p. 42). The plant was to be finished by mid-1996—two and one-half years later at a cost of $92.6 million.

METHANOL

Methanol is the alcohol of methane: methyl alcohol CH_3OH.

Methane gas is converted into synthesis gas, which is a mixture of carbon monoxide and hydrogen gas, and then reassembled into methanol. It is used as a feedstock in the petrochemical industry and can be used as an automotive fuel directly or indirectly. The indirect use is as a feedstock for methyl tertiary butyl ether (MTBE).

The capital cost for a world-class, 2,000-ton/day methanol plant requiring up to 60 MMCFD inlet gas can range from $250 to $300 million.

FERTILIZER

Natural gas is the feedstock for manufacture of ammonia, which is the primary feedstock for fertilizer known as urea.

In early 1994 plans were announced for a world-class ammonia urea complex at Gresik, East Java. Costs were estimated at $242 million for the 1,350 metric ton/day ammonia–1,400 tons/day urea plant (*Petromin Magazine,* March 1994). This particular plant would probably not be considered to be a balanced plant. Only about 800 tons/day of ammonia would be required to manufacture 1,400 tons/day of urea. But ammonia has uses other than as a feedstock for fertilizer. The plant example outlined in Table 6–2 is a balanced plant requiring 80 MMCFD feed gas. The plant is assumed to produce 1,000 tons/day of ammonia and 1,750 tons/day of urea.

LIQUIFIED NATURAL GAS (LNG)

LNG is liquified methane and ethane. The problem that faces LNG development is primarily the huge up-front capital cost of building a full-range LNG chain from gas field development to liquefaction, transportation, and receiving terminals. The liquifying temperature for LNG is –162°C. As a result, the processing, storage, and transportation are quite expensive.

A new LNG plant can easily cost $2 billion, with each expansion train of an existing facility costing $400–$600 million and taking from 24 to 36 months to complete. Existing facilities have a huge cost advantage over grassroots construction. The cost for expansion is nearly half. In March 1994 the Badak Train-F 2.3 MMT/Y capacity LNG plant at Bontang, East Kalimantan, came onstream. The cost was $522 million, including infrastructure support, housing, and roads. Interest during construction added another $177 million, and the total cost came to $699 million. Construction on Train-G also at Bontong was due to begin in April 1994 at an estimated cost of $580 million.

Table 6–2 summarizes vital statistics for world class gas development options. The objective here is to give an indication of the thresholds and boundary conditions that govern or influence international gas projects. Every situation is different, and cost estimates that are not site specific are notoriously inaccurate. The difference in construction cost can vary by over 100% from one location to another, depending on many factors. This is illustrated to some extent by the difference between grass roots construction and expansion with LNG facilities—a 4:1 difference. There is another important difference. Most upstream project costs that are quoted do not include interest during construction. Most gas development and downstream project costs quoted in the press do include interest during construction. There is a similar difference with operating costs. Upstream operating cost quotes do not include DD&A. When those in the downstream end of the industry quote operating costs, they ordinarily include DD&A. They may also include feedstock costs, which can make a big difference. Regardless of these heavy qualifiers, it was felt that at least an attempt to give ballpark cost figures would be better than nothing. Interest during construction and DD&A are not included in the cost estimates summarized in Table 6–2. Neither are feedstock costs included in operating costs.

There are additional costs that have not been included. Many options require an export terminal, and export products require tankers. The cost for a terminal can range from as low as $200

Table 6–2

	GAS DEVELOPMENT OPTIONS					
	Gas Sales	LPG Plant	Gas Cycling	Methanol	Fertilizer Ammonia/Urea	LNG Plant
Product	Natural Gas and Condensate	LPGs and Condensate + Gas	Condensate Gas is reinjected	Methanol Hydrogen	Granulated Urea	Liquified Methane & Ethane
Threshold Field Size (BCF)	Wide range 50–200	300–400	250–400	500	600	4,000
Minimum Feed Gas (MMCFD)	10–20 MMCFD	60–80 MMCFD	40–75 MMCFD	60 MMCFD	80 MMCFD	600 MMCFD 300/train
% Produced/Yr Project Life	7% or more 5–12 Years	5%–10% 10–20 Years	7%/Year 13 Years	5%/Year 20 Years	5%/Year 20 Years	4%/Year 25 Years
Capacity	20 MMCFD	60 MMCFD 4,000 BCPD	30 MMCFD 1,000+ BCPD	2,000 tons/day	1,750 tons/day	11,000 tons/day 4.3 MM tons/year
Market Requirements	Local Pipeline	Local & Export	Local	Export Ship	Local Truck	Export LNG Tanker
Plant Location	No Plant	Local or Port City	Local	Port City	Local	Port City
Required Capital Cost $MM	$5–$10+	$50–$60	$75–$100	$250–$300	$150–$200	$2.5–$3,000 Grassroots
Lead Time	4–24 Months	3+ Years	2–3+ Years	4 Years	5 Years	7–10 Years
Construction		2+ Years	2+ Years	3 Years	3 Years	3+ Years
Annual Op. Costs	$4 MM	$8 MM	$15 MM	$30 MM	$35 MM	$80 MM
Current (1994) Product Prices $/MCF Equivalent	50¢–$2.50/MCF 50¢–$2.50	$110–$130/ton $3.00/MCF	$10–$11/bbl Condensate	$125–$140/ton	$160–$180/ton	$200/ton
Threshold product price to build now	$1.00/MCF	$160/ton	$12/bbl	$180+/ton	$180+/ton	$220/ton
		Economics strongly affected by gas price	Liquid/Gas yield needs to be above 30–40 bbls per MMCF			Minimum based on 2 process trains

Long-term minimum take-or-pay contracts required to start |

[1]Excluding Feedstock

129

million to over $500 million for deepwater port facilities capable of handling large tankers. Also, extended distances from the field to the plant or port facilities will increase pipeline costs. Extra large export distances can add even more. For example, a round trip from Mobil's Arun LNG complex in Indonesia to Japan is about 15 days. A typical cargo for an LNG tanker with 125,000 m^3 storage is 57,000 tons, equivalent to about 2.7 BCF of liquified gas. Assuming it takes 20 cargoes per year, a tanker can transport about 1.14 million tons per year—150 MMCFD equivalent. The two-train world class LNG facility in Table 6–2 producing 4.3 million tons per year would need four tankers at well over $100 million each. LNG is big business.

7

GLOBAL MARKET FOR EXPLORATION ACREAGE

Governments are becoming increasingly aware of their position in the global market for exploration acreage or rehabilitation projects. The market for drilling funds and technology is supercompetitive and getting sophisticated.

Most countries developing petroleum fiscal systems are choosing the PSC. It is an obvious trend that began in the 1960s. The financial results could be the same with a concessionary agreement, depending upon the aggregate level of taxation. Philosophical and political considerations give the advantage to the PSC.

There are still promising sedimentary basins on this planet that are virtually unexplored. These basins are more geologically complex, more remote than the established provinces, or they are dominated by a harsh environment. Yet even basins with

131

established drilling and production history have a long way to go in terms of exploration and development maturity by comparison to the United States.

Many regions are at a stage that existed 20–40 years ago in the United States. The total wells drilled in the United States now exceeds 3.2 million. This is two orders of magnitude greater than the 30,000-odd wells that have been drilled in the Asia Pacific region (excluding China). Seismic data acquisition in the United States is about 10 times greater. There have been roughly 6.2 million mi of seismic data acquired in the United States compared to 650 thousand mi of data acquired in the Asia Pacific.

Four times as many wells have been drilled in the United States as in Russia. If the industry drilled 50,000 wells per year in Russia, it would take 40 years to catch up to where the United States is *now* just in terms of the number of wells drilled. Variations in maturity like this account for much of the shift in focus from the United States to the international sector. Well productivity outside the United States is nearly 20 times that of a typical 12-B/D U.S. oil well.

Of the nearly 600,000 producing oil wells in the United States, over two-thirds are stripper wells—producing less than 10 BOPD. The average production rate for these 400,000-odd stripper wells is just over 2 BOPD. There is still geological potential in areas of the United States, but most of these areas are off-limits to the oil industry.

One perspective on the geological scale offered overseas can be seen in the kind of production rates found. Figure 7–1 summarizes oilwell production rates of various countries and compares them to production statistics from the United States. This figure also illustrates the advanced maturity in the United States.

Many countries have good exploration potential, and they know how to compete for Western capital. Fiscal terms in most of these countries are tougher than those in the United States, but not in relation to the fundamentals of political risk, geology, and

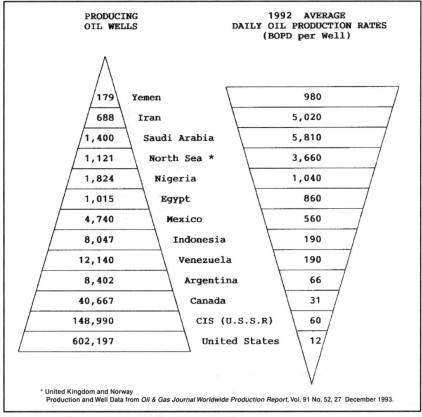

PRODUCING OIL WELLS		1992 AVERAGE DAILY OIL PRODUCTION RATES (BOPD per Well)
179	Yemen	980
688	Iran	5,020
1,400	Saudi Arabia	5,810
1,121	North Sea *	3,660
1,824	Nigeria	1,040
1,015	Egypt	860
4,740	Mexico	560
8,047	Indonesia	190
12,140	Venezuela	190
8,402	Argentina	66
40,667	Canada	31
148,990	CIS (U.S.S.R)	60
602,197	United States	12

* United Kingdom and Norway
Production and Well Data from *Oil & Gas Journal Worldwide Production Report*, Vol. 91 No. 52, 27 December 1993.

Figure 7–1 Comparison of oilwell production rates

economic potential. These governments are aware of their position in the global market and understand fairly well what the market can bear. They are watching closely the hyperactivity in Eastern Europe and the former Soviet Union. The growing competition for drilling funds and Western technology should benefit companies in the international exploration business. There are a lot of different opportunities in the international sector—something for nearly everyone. It is a slightly different game, but rocks are rocks and taxes are taxes.

During the 1990s activity in the former Soviet Union and Eastern Europe took on the aspect of a modern-day gold rush.

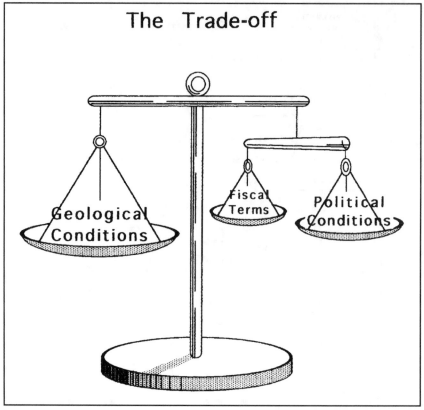

Figure 7–2

During the first quarter of 1991 the Soviet oil output hit a 15-year low, and the region is desperate for Western capital and technology. Major international oil companies and companies that have never ventured overseas are flocking to this region. Russia promises to be a hub of activity, if it can live up to its promise. To do this it must create a competitive investment climate and overcome the curse of political instability and the famous Russian bureaucracy.

Everything must balance. This is shown in Figure 7–2. If the balance is too far in favor of oil companies' interests, then perhaps in the short term there may be a flourish of exploration activity. But in the long run, it is unlikely that an unbalanced situation will persist. Too often fiscal adjustments follow successful exploration efforts.

Contract negotiations must seek the appropriate balance.

Part of negotiations is knowing the market. Obtaining terms that are good in the global context is one thing, but nobody wants to negotiate the worst terms in a country, even if they are relatively good terms. Nearly half of the equation in negotiations is to secure terms that will be acceptable to a potential partner. Knowing the market and what kind of terms are realistic depends on a region's prospectivity.

PROSPECTIVITY

The most obvious geotechnical aspects of exploration are field sizes and success ratios. Targets must be large, and just how large they must be depends on costs and fiscal terms. Sometimes countries are compared on the basis of barrels of oil discovered per wildcat. The question is, Will past statistics provide some insight? Just how meaningful historical information is in regard to success ratios and the size of discoveries depends on the maturity of an area and other factors. If all the major structures have been drilled, then the past is certainly not the key to the future.

Costs are critical. With maturity and infrastructure development in a given province, costs often come down. This is particularly true with exploration and transportation costs, which are highest for the first field developments in an area.

Well productivity is a dominant factor in development costs. This is particularly true with offshore developments and increasing water depth. Wells that produce 500 BOPD in the United States are rare and exciting, but in the international arena, particularly in deepwater or frontier areas, they may well be marginal or even uneconomic.

While most fiscal systems in the international arena are tougher than in the United States, the more favorable geological potential usually compensates. The chance of finding over 100 MMBBLS is realistic in Latin America, Southeast Asia, and West Africa but unlikely in most of the United States. The exceptions in the United States are the deep water of the Gulf of Mexico, offshore California, and Alaska. Political risks are also normally greater, but again, if there is sufficient geological potential and the fiscal terms are appropriate, there may be opportunity.

POLITICAL RISK

One of the major considerations in international operations is the element of country risk. How can companies gauge the attendant risks of doing business outside their own country? It is not easy.

Nationalization or expropriation of assets once loomed foremost as the greatest risk that the industry faced overseas. It is still the worst nightmare for many a company or negotiator, as shown in Illustration 7–1. However, the scrutiny of the expanding global economy and the interlocking financial markets make this kind of action much more unlikely than in the past. The international business and financial community can impose severe penalties on a government that expropriates the assets of an expatriate company, regardless of whether or not the company receives compensation. It is not a decision that is taken lightly.

Expropriation is not illegal in the eyes of international law as long as it is done in the best interests of the country and the companies involved are compensated. These are the most common aspects of legal expropriation. But adequate compensation is seldom the hallmark of these kind of government actions. However, governments know that even a hint of nationalization, expropriation, or confiscation can send ugly signals. It can take years to reassure potential investors. While confiscation of assets is a risk, it is often overshadowed by other factors. More realistic risks include such things as *creeping nationalization* through expanding taxes, progressive labor legislation, or price controls.

POLITICAL STABILITY

One of the more difficult aspects of conducting business in another country is the element of never-ending rule changes.

Illustration 7–1 A negotiators worst nightmare—expropriation

Policy shifts constitute the most prevalent and immediate risks that confront the industry. These include changes in government and fluctuating tax laws. In some countries the rate of change is excessive. Democracies, for example, have a habit of making changes that affect the business community nearly as often as elections are held. In terms of volatility and unpredictable tax law changes, the U.S. record is not the best. The United States has shown an eagerness to impose excess (windfall) profits taxes, yet the government is not nearly as responsive as other countries are to make compromises when adverse conditions face the industry.

Contract flexibility can add an element of stability. Theoretically governments are less likely to change the rules if fiscal or contract terms themselves are responsive.

Figure 7–3 shows where political risk fits into the risk analysis framework. The best way to evaluate or quantify aspects of political risk or any other kind of risk is with expected value theory (EMV theory). Expected value theory is the central theory, and possible scenarios or outcomes are represented by their financial impact through discounted cash flow analysis. The fiscal terms of a country and the political stability have a huge impact on the country's prospectivity.

Aside from the geological dimension, almost all of the elements that affect the investment decision are or can be influenced by the government and the nature of the business climate offered to the industry. The investment decision is also influenced by *utility theory* and *gambler's ruin theory*, both of which are strongly influenced by the estimated probability of success and the nature of the perceived risks.

GAMBLER'S RUIN THEORY

Gambler's ruin occurs when a risk-taker with a limited amount of funds goes broke through a continuous string of failures. With any kind of drilling budget there is such a risk, but it can be reduced through diversification.

The concept of gambler's ruin flows from the same estimates of risk used in expected value theory. The estimate of the probability of success is used to help determine the maximum level of capital exposure under specific *confidence level* criteria. The objective for management is to stay in the drilling game long enough for the odds to work as they should.

Another way of looking at it would be to ask the question, What would be a statistically significant sampling size (at a confidence level of 95%), given the nature of the play?

The concept of *gambler's ruin* is used for avoiding a successive string of failures that would exhaust a drilling budget. For example, at least one success must be achieved. It is a rather fancy concept for a well-known adage: "Don't put all your eggs in one basket!"

The gambler's ruin algorithms are outlined as follows:

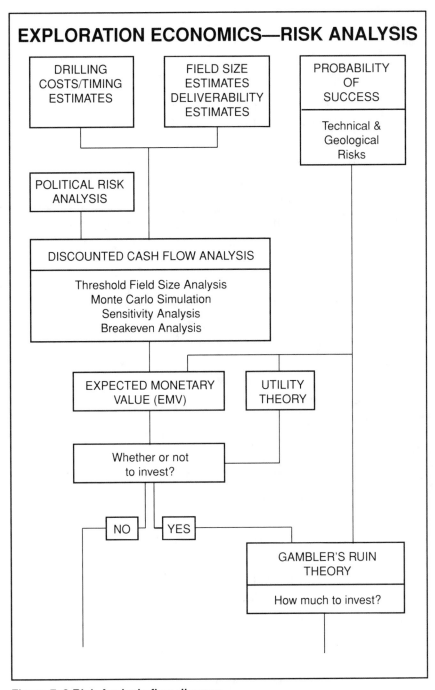

Figure 7–3 Risk Analysis flow diagram

Probability of at least one success	= 1	–	Probability of all failures

Assume that a confidence level of 95% is desired, and the probability of success is estimated at 15%. How many wells must be drilled to be 95% confident of having at least one success? At least 18. With 18 wells there is still roughly a 5% chance of having all dry holes. The maximum percentage of the total budget that can be spent is 5.4%.

Two investors can participate in the same drilling venture: one gambling and the other exercising sound investment strategy. The difference is in the level of exposure.

GAMBLER'S RUIN ALGORITHMS

$.95 = 1 - (1-sp)_n$
where:

.95	=	Desired confidence level
sp	=	Probability of success
$(1-p)$	=	Probability of failure
n	=	Number of trials (exploratory wells)

Solving for n:

$$n = \frac{\log (0.05)}{\log (1 - sp)}$$

where:

0.05	=	1 – Desired confidence level Probability of all failures
sp	=	Probability of success
$(1-p)$	=	Probability of failure
n	=	Number of trials (exploratory wells)

UTILITY THEORY

Human and corporate behaviors manage to carry on, whether or not people know or care that there is an exotic name that applies to their actions. Utility theory is also known as preference theory. It describes to a large extent why people will happily stick a quarter into a slot machine in Las Vegas even though the odds are squarely against them. The expected value of that sort of action is negative—always. This is called gambling. But quarters have almost no utility. When it comes to risking a few million dollars on an exploratory well, even expected value theory is not enough. Expected monetary value theory explains what people should do and what the boundary conditions are. It does not explain behavior. If the expected value of a potential investment opportunity is positive, then it is worthy of consideration. Just how positive must an expected value be? The standard industry two-outcome EMV model is used once again in Figure 7–4 to illustrate the essence of utility theory. As before, the risk capital is $15 million. The possible reward in this two-outcome model has a value of just over $100 million. These points define the EMV curve. Staying below the curve results in positive expected values. For example, the expected value is equal to $20 million with an estimated probability of success of 30%. If management were bidding on this project, the bonus bid would have to be less than $20 million. According to the utility curve, the bid would have to be less than $2 million.

The EMV break-even success ratio is close to 13%, but the utility break-even success ratio is over 25%. Utility curves and EMV curves approach each other at the end points. If management was convinced that the $15 million well had 0% chance of success, then it would be worth a negative $15 million. No company would drill such a well—unless someone paid the company $15 million to do it. Drilling contractors do it all the time. On the other hand, if management were convinced that this drilling opportunity had a 99% or 100% chance of success, the expected value and utility values converge. It would be like buying production or proved undeveloped reserves. It is in the middle ground where the curves diverge. Companies have different risk profiles or levels of risk aversion. Under the same set of assumptions, the EMV curves would look the same, but the utility curves would not. Determining a company's utility curve with a drilling venture like this is complicated and gets academic. It is not done often. Sometimes the whole subject is simply marked down as gut feel and left at that.

It was pointed out in Chapter 2 that the endpoint values represent the result of discounted cash flow analysis, discounted at the corporate cost of capital. The margin between the utility curve and the EMV curve is sometimes viewed as the minimum *risk premium*. This is one reason why negotiations get sticky on the subject of profitability. For drilling ventures like this that are successful, the rewards are spectacular, but only within the narrow context of the discovery well—which requires ignoring any associated dry holes. Once a discovery is made, it is hard to speak in terms of chance of success. This is particularly true from a government's viewpoint because it may not be aware or may not care that the company has drilled five dry holes before the discovery. Utility theory and expected value theory suddenly become useless once a discovery is made. And unfortunately, the bargaining power of the company suffers accordingly.

BARGAINING POWER

While geological risks begin to diminish after discovery, political and financial risks intensify. Bargaining power of the contractor begins to diminish also. The essence of bargaining is power, and the relative strength of bargaining positions shifts during the cycle of petroleum exploration and development. This shift is shown in Figure 7–5.

Once a resource project becomes commercial, bargaining power really begins to shift. The large investments for the devel-

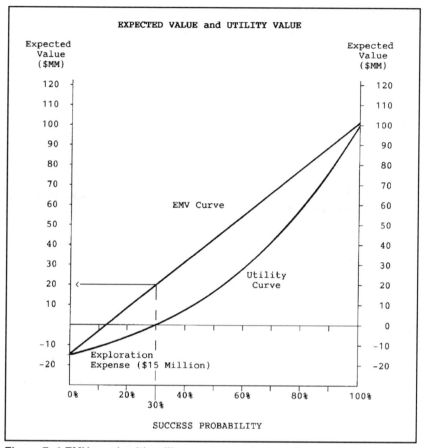

Figure 7–4 EMV graph with utility curve superimposed

opment phase of petroleum operations start out as a source of strength for the contractor. By the time production commences, capital investment is a sunk cost, and facilities installed in a foreign country can represent a significant source of vulnerability to the contractor.

Payout is another milestone where the shift increases in favor of the government. Once the contractor has reached payout, profitability can appear to be excessive to some governments. The concepts of profitability and a *fair return on investment* start to be reviewed at this point. The subject of exploration economics and

143

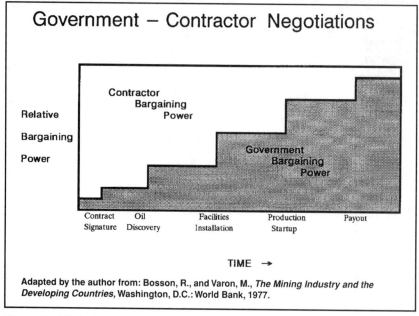

Figure 7–5 Relative bargaining power

the substantial risks taken by the petroleum industry are hard to sell. The contractor's position has changed.

Outcomes at all stages in the life cycle of a petroleum license are determined by bargaining power. The relative bargaining power of the oil company is greatest in the early stages prior to contract signing and after that, before discovery. There are natural milestones along the evolutionary path of ultimate bargaining power, but the trend clearly shifts power over to the state as the cycle progresses.

OPERATIONAL RISKS

The partnership between a government and an outside oil company, whether formal or informal, goes beyond written contract terms and conditions. Good working relationships can make a big difference.

The operational risks are some of the real, tangible risks associated with doing business in a foreign country. If it takes a week to obtain a visa, then imagine what it might be like trying to import equipment, supplies, and personnel.

LANDLOCKED COUNTRIES

One of the worst negotiating positions occurs with landlocked countries. Dealing with one government is enough of a challenge. Dealing with two is more than twice as difficult. The dynamics multiply. It seems that neighboring countries never seem to see eye to eye, especially when an oil company wants to cross one of them with a pipeline. This dilemma can tax even the most seasoned and experienced negotiators.

QUANTIFYING POLITICAL RISK

One of the most frequently asked questions regarding political risk is, How do you *quantify* political risk? The best way is to use the expected monetary value theory described earlier in this book. The following example in Table 7–1 shows the possible scenarios envisioned for a hypothetical exploration venture. It is assumed that the exploration effort could be subject to varying degrees of political unrest, as well as possible expropriation. The table shows the relative weighting attached to the various possible outcomes. The project appears to be worth consideration even though there is an acknowledged risk of expropriation as outlined in the EMV model.

In this example the expected value is $51 million, in spite of the obvious risks. The model accounts for an 8% chance of expropriation in addition to a substantial risk associated with political unrest in the country. The table is another representation of EMV theory, but allows more than two possible outcomes. It is a tabular format for what is known as a *decision tree*. In many respects, whether or not management explicitly estimates probabilities or models the economic results of possible scenarios, the risks and rewards are weighed and balanced. Sometimes it is called gut feel, intuition, or experience. While this example has a positive expected value, many managers would prefer a healthier climate.

145

Table 7-1

Quantifying Political Risk

Possible Outcome	Present Value ($MM)	Probability (%)	Expected Monetary Value ($MM)
Unsuccessful Exploration Effort: $50 MM Work Program	−$50	70%	−$35
Large Discovery Moderate political unrest	$700	5%	$35
Large Discovery Substantial political unrest	500	5%	25
Moderate Discovery Moderate political unrest	350	6%	21
Moderate Discovery Substantial political unrest	250	6%	15
Large Discovery Assets expropriated during development work	−150	4%	−6
Moderate Discovery Assets expropriated during development work	−100	4%	−4
		100%	$51

Risk aversion/utility values come into play. As mentioned earlier, EMV theory is only part of the equation.

It is one thing to quantify political risk, regardless of how it is done. It is quite another thing to protect against it. The first thing that comes to mind is insurance. This is one way to transfer risk as outlined below, which is one of many ways to manage risk.

INSURANCE

A small percentage of foreign investment is insured. However, in some countries eligible for foreign assistance and insurance coverage the number of projects insured can be quite high. Many risky countries are eligible for various types of political risk insurance covered through international organizations, but some are not.

Insurance coverage is available for virtually any aspect of country/political risk. A couple of the more prominent include the Overseas Private Investment Corporation (OPIC) and the Multilateral Investment Guarantee Agency (MIGA). Both of these entities require that they be notified before any commitment is made. If a contractor desires coverage for any aspect of political risk through these agencies, the contractor must inform them early on. Since they are involved from the inception of an investment, they are interested in the contractor's means of risk management.

RISK MANAGEMENT INCLUDES:

- **Risk Avoidance**
  Forgoing an opportunity because risk of loss is too great
- **Loss Prevention**
  Reduction of loss frequency
- **Risk Retention**
  Not all risks are avoidable
  Establishment of funded reserves to offset losses
- **Risk Transfer**
  Similar to loss prevention: spreading risk through
  farmouts and joint ventures
- **Insurance**
  Variation of risk transfer

Of all the risks that exist, the greatest concern centers on the specter of expropriation. In the past there have been spectacular losses from nationalization, expropriation, and confiscation of industry assets. The following outlines provide basic rules for reducing risk of expropriation or at least minimizing losses.

Reducing Expropriation Risk

- Keep a low profile
- Maintain close relationships with the government
- Avoid layoffs (particularly of nationals)
- Avoid geographical overconcentration
- Utilize local industries and national personnel

Minimizing Expropriation Losses

(in the face of impending expropriation)
- Stop new investment and shipments
- Cut back inventories and cash
- Pay down debts to home country suppliers
- Borrow heavily from local sources

OPIC is the best known agency for insuring aspects of political risk. OPIC was created as an independent U.S. government agency through the Foreign Assistance Act of 1969. OPIC is limited by statute to insure projects in less developed countries (LDCs). However, the principal initiative of OPIC was to meet the needs of U.S. investors in foreign petroleum and mining projects. The primary kinds of coverage that OPIC provides are for:

- **Inconvertibility**
  (inability to convert payment in local currency received as income—does not cover devaluation)
- **Expropriation**
  (includes confiscation and nationalization including creeping expropriation)
- **War**
  (includes revolution and insurrection)

OPIC insures only net unrecovered costs of projects. This is one reason why the larger companies in particular do not use

OPIC insurance that much. The risk of substantial contract revisions in the early exploration stage of a contract are just not as great as they are after a discovery is made, or after the company has recouped most or all of its costs.

As a general rule, war risk premiums for OPIC political risk insurance are higher than expropriation risk premiums. OPIC insurance premiums for example would range from 0.3 to up to 1.5% of coverage per year with 90%–95% indemnification of the amount insured. By comparison, a premium for earthquake coverage might be on the order of 0.1% of coverage, depending upon the region. OPIC coverage for war, revolution, or insurrection requires that the disruption last for more than six months. OPIC evaluation criteria include the nature of the project and the arrangement with the host government, including incentives and regulations. Some of the most important considerations are the effect on the economy, employment, wages, local suppliers, downstream industry, and the environment.

MIGA was established by the World Bank in 1985. It was established to promote private investment in the member countries, including Switzerland, and the insured must be citizens of member countries. The insurance coverage is similar to that offered by OPIC.

The United States Export-Import Bank (Eximbank) also provides some means of political risk insurance for U.S. manufactured equipment exported overseas.

Private insurance is also available, but premiums are generally higher. There is more flexibility with private insurance and confidentiality is better. Once a contract is issued with OPIC or MIGA it becomes public information. Private organizations cover virtually everything that OPIC or MIGA cover.

The key to success in the international industry is to be able to evaluate and to manage the attendant unavoidable risks and ensure that they are more than outweighed by the potential rewards. The industry is a risk-taking industry, but avoiding undue risks and minimizing risk is a winning strategy.

8

PRODUCTION SHARING CONTRACT OUTLINE

Examination of a PSC, service contract, or even a concessionary arrangement must consider the greater regulatory and legal context. Nearly every country has a legal and legislative framework within which all agreements reside. Figure 8–1 illustrates the legal and contractual foundation for petroleum agreements. It starts with the constitution. Not all countries have the same framework shown in Table 8–1. The foundation is the nation's constitution. From the constitution flows the taxing and legislative authority that governs petroleum legislation and outlines the authority and boundary conditions for relationships with foreign companies.

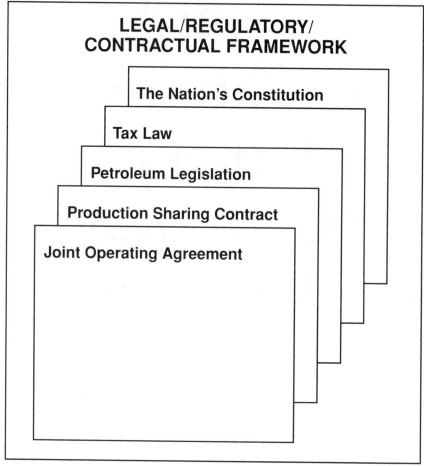

Figure 8–1 Legal/Regulatory/Contractual Framework

THE CONSTITUTION

The constitution of a country is the foundation upon which all else must stand. Petroleum legislation and individual contracts cannot contradict constitutional law. For example, the Venezuelan Constitution required that any contract disputes involving the public interest be resolved exclusively in Venezuelan courts. With that in mind, it would be difficult or foolish to draft an arbitration clause in a petroleum contract that places the venue for arbitration outside of Venezuela. The constitution dictates the lines of authority for the various aspects that govern the relationship between the contractor and the state.

TAX LAW

Sometimes the taxes pertaining to the petroleum industry or industry in general are found under a separate set of laws. The tax liabilities that the contractor is subject to may be included in the contract by reference to the pertinent tax law, or simply by mention of the fact that taxes must be paid. The Indonesian PSC does not mention the 35% income tax, nor does it mention directly the additional 20% withholding tax. The reference is oblique but not intended to mislead. The taxes are simply defined by a higher authority. The Indonesian production sharing contract clause that deals with taxes effectively falls under the Rights and Obligations of the Contractor:

Contractor Shall:
be subject to and pay to the Government of the Republic of Indonesia the Income Tax and the final tax on profit after tax deduction imposed on it pursuant to the Indonesian Tax Law and its implementing Regulations. CONTRACTOR shall comply with the requirements of the law in particular with respect to filing of returns, assessment of tax and keeping and showing of books and records.

PETROLEUM LEGISLATION

Petroleum legislation in many countries can be rather archaic and can contain many irrelevant procedures. Pertinent legislation governed by the constitution may include specific petroleum legislation that authorizes the national oil company or oil ministry to negotiate certain aspects of agreements between the state and foreign companies. Some governments have no petroleum legislation, and the arrangement between the contractor and the state may be embodied totally under the terms of a contract. Some such contracts even dictate the taxes to be paid.

PRODUCTION SHARING CONTRACT

The PSC is the official link between the government and contractor in those countries that use this system. For those countries that do not, a concessionary agreement is used, but it will cover most of the same aspects that are negotiated in a PSC. In either case, if there is government participation, a joint operating agreement may also be required.

ENVIRONMENTAL/CONSERVATION REGULATIONS

Environmental laws are being drafted at a furious pace, and if there is no written law, then it is likely there soon will be. Environmental regulations are important and becoming even more so. Even if there is not specific contract language that addresses this issue, many feel that the typical contract clauses that require the contractor to exercise "prudent oilfield practices" include the obligation to protect the environment.

Table 8–1 shows some of the diversity of legal frameworks found in the petroleum industry.

PSC COMPONENTS

The following outline contains typical sections that are covered in production sharing contracts between host governments and foreign oil companies. There are many variations, but the following descriptions are fairly standard. Furthermore, the issues discussed here are relevant to other fiscal arrangements as well, whether they are embodied in a PSC or petroleum law or a license agreement.

Table 8–1

Examples of Legal/Regulatory Frameworks

	Type of Government*	Constitution*	Legal System*	Petroleum Legislation	Authorized State Agency
Papua New Guinea	Westminister Style Parliamentary Democracy	9/16/1975	English Common Law	Petroleum Act 1977 - 4/1978 Chapter 198 Income Tax Act 1978 Chapter 110	Department of Minerals and Energy
Malaysia	Constitutional Monarchy	8/31/1957 Amended 1963	English Common Law	Petroleum Developments Acts 1974/5	Petronas Nat. Oil Co. Carigali Exploration Arm
Nigeria	Military Government Since 1983	10/1/1979 Amended 2/84 Revised 1989	English Common Law Islamic Law and Tribal Law	Petroleum Drilling & Prod. Regs. of 1969 as Amended	Nigerian Nat. Oil Co. (NNOC)
Colombia	Republic — Executive Branch Dominates	7/5/1991	Based upon Spanish Law	Decree 2310 of 1974 as regulated by Decree 743 of 1975 Resolution 000058 (prices)	Ecopetrol Nat. Oil Co.
Tunisia	Republic	6/1/1959	Based upon French Civil Law and Islamic Law	Decree Law Nº 85-9 14/9/85 Modified by Law Nº 87-9 6/2/87 and Decree Law Nº 90-55 18/6/90	Minister for Energy and Mines
Vietnam	Communist State	12/18/1990	Based upon Communist Legal Theory and French Civil Law	Law of 29 Dec. 1987 on Foreign Investments	Petrovietnam Nat. Oil Co.
Ivory Coast Côte d'Ivoire	Republic Multiparty Presidential Regime	11/3/1960	Based upon French Civil Law and customary Law	Law of 64-249 7/1964 amended by Decree 64-96 3&4/1965 Law 70-489 of 3 August, 1970	PETROCI Nat. Oil Co.
Qatar	Traditional Monarchy	Provisional Constitution enacted 4/2/1970	Discretionary system of law controlled by the Amir and Islamic Law	No Petroleum Law	Qatar Petroleum Producing Co. QPPC Nat. Oil Co.

*From: The World Factbook 1991, Central Intelligence Agency

PSC Outline

Article	Title
I	General Scope
II	Definitions
III	Purchase of Data
IV	Duration, Relinquishment, and Surrender
V	Work Programs and Expenditures
VI	Production Areas
VII	Signature and Production Bonuses
VIII	Rights and Obligations of National Oil Company
IX	Rights and Obligations of Contractor
X	Valuation of Petroleum
XI	Payments
XII	Recovery of Operating Costs and Net Sales Proceeds Allocation
XIII	Employment and Training of Local Personnel
XIV	Title to Equipment
XV	Ownership Transfer
XVI	Books and Accounts
XVII	Procurement
XVIII	Joint Operating Agreement
XIX	Force Majeure
XX	Arbitration
XXI	Insurance
XXII	Termination
XXIII	Entire Contract and Modification
Exhibit A	Description of Contract Area
Exhibit B	Map of Contract Area
Exhibit C	Accounting Procedure
Exhibit D	Management Procedure

General Scope

The first section of the contract outlines who the various parties to the contract are. This will include the foreign oil company, or companies if there are more than one, and the national oil company or the agency acting on behalf of the host nation. The oil companies are identified and thereafter referred to as the "contractor." The host government national oil company or agency is also identified and then subsequently referred to by initials or a short name.

DEFINITIONS

This is a standard contract section that defines specifically technical and financial terms to promote a common understanding and source of reference. The definitions can be fairly straightforward and can provide an important foundation for mutual understanding of the contract. However, terms such as *force majeure, effective date, commercial discovery, production period, wildcat well, exploration well,* and *appraisal well* often seem quite simple until there is a problem. Then the contract is hauled out, and any differences or confusion between the definitions and any corresponding language within the body of the contract will be the focus of much discussion or perhaps arbitration.

PURCHASE OF DATA

This clause specifies that a data package consisting of geological and geophysical information must be purchased. The purchase price is stated, and a brief description of the data package is included. This clause will often stipulate that the purchase price for the data package will not be recoverable under the cost recovery provisions of the contract.

Costs for data packages range from $10,000 to upwards of $75,000 and more in some countries. Data package quality can sometimes be fairly good.

Sometimes the data package purchase requirement precedes the contract signature or negotiations, and it may be stipulated that the data be acquired before beginning negotiations. The price of the packages in those cases is not as much an indication of data quality, perhaps, as it is a means of screening applicants.

DURATION, RELINQUISHMENT, AND SURRENDER

Sometimes these aspects of the contract are placed under separate clauses, but they are strongly interrelated. Duration language includes a description of the length of the exploration and production phases of the contract and the terms under which extensions may perhaps be granted and the duration of the extensions.

The primary exploration phase ranges from two to more than six years without extensions. At the end of the exploration period, the contractor is usually required to surrender/relinquish a portion of the contract area. This obligation may sometimes be waived if an extension of the exploration period is granted. Contracts specify conditions under which extensions may be granted, which may include a bonus, technical justification of some sort, or additional work. Extensions will allow an additional one to three years, depending upon circumstances.

This clause requires careful consideration under ordinary circumstances. But it is particularly important under extreme climatic conditions, such as above the 50th parallel where spring breakup can shut down operations or where weather windows limit the amount of time each year during which companies can operate. Monsoons, winter conditions, spring breakup, and iceberg seasons are a few examples. Not only do these kinds of conditions limit available time, but coordinating operations between windows becomes critical. What may take a year to do in south Texas may require two years or more in a rice paddy or three years just south of permafrost.

Some contracts outline a *delineation* phase after discovery, with an additional well commitment. Delineation periods are much like exploration extensions and range from one to two years. The importance of this phase is that the contractor's delineation drilling costs are sometimes not reimbursed upon government back-in. When predevelopment phase costs are not reimbursed, or if the government is carried through this phase, the importance of obtaining sufficient information or test results is critical to prove commerciality as soon as possible.

The development phase of a contract, often referred to as the *exploitation* phase, can last from 20 to 30 years or more. Anything less than 20 years is unusual and not recommended.

Normally there is a distinction between the exploration and development phases of a contract. A typical relinquishment clause may stipulate that after a three-year exploration phase, the contractor must relinquish or surrender 25% of the acreage. Sometimes, with inconclusive exploratory drilling, extensions may be granted for delineation drilling and development feasibility studies. Area reduction clauses are standard, and the most common percentages that are quoted refer to percentage of "original" area. But, like many other aspects of this business, this is not an absolute. The term *backaway* is sometimes used for the contractor's option of total relinquishment after the exploration phase if all work commitments have been fulfilled.

WORK PROGRAMS AND EXPENDITURES

This is one of the more important sections of the contract. The work program is ordinarily specified in terms of kilometers of seismic data to be acquired and the number of wells to be drilled.

Such things as shooting and processing parameters for the seismic and minimum footage requirement for the drilling obligation may be included. Sometimes a minimum expenditure level regardless of the number of wells is required in the work commitment. For example, a contractor may agree to a work commitment of at least $20 million or three wells, whichever is greater. Contracts may also require that wells be spudded within a certain period, such as within 12 to 24 months of contract signing.

Problems here can occur with well classification or definition. For example, suppose an operator drilled a successful exploratory well as the first of a two-well commitment and drilled the second well 3 mi away to find that the accumulation encountered in the second well was actually an extension of that found by the first. The government may argue that the well was a delineation well on the first discovery, and that another exploratory well was required to fulfill the obligation.

Another example: Suppose two wells were planned for separate fault blocks, and the first well was dry. The operator decides to back-up and sidetrack the first well and drill directionally to test the second fault block. The objective of testing both fault blocks was met, but the government asks, "When will the *second* well be spudded?"

In most exploratory efforts, the relationship between contractor and government will not get beyond the work program stage because most exploration efforts are unsuccessful. The structuring and wording of the work program is important to both contractor and government.

In most contracts this section also outlines the authority of the host government to review all budget proposals on a periodic basis. The host government's national oil company or oil ministry will have the right to make suggestions and to propose revisions to any work plan and budget. However, the contractor normally has the authority to make final decisions on matters concerning the work program once the contract has been negotiated.

PRODUCTION AREAS

Some contracts specify that once a discovery is made, the productive limits of the field must be identified and mapped. The reason for this, in some cases, is that costs associated with development of a discovery may be treated differently than costs associated with exploration efforts. Furthermore, when the time for relinquishment arrives, the productive areas are exempt. In Chile the government grants a 5-km *protective halo* around fields as part of the definition of production area. In other countries specific contract terms (such as the Indonesian DMO) apply on a field-by-field basis, so the determination of productive area is critical.

SIGNATURE AND PRODUCTION BONUSES

The signature or signing bonus is well known and highly unpopular with the oil industry. The signature bonus is an artifact of competitive bidding, but can easily be part of negotiated deals. The signature bonus is simply a payment that occurs at or as a function of contract signing.

Some contracts will specify that added bonuses be paid at various milestones such as threshold production levels or cumulative production landmarks. For example, the contract may stipulate in this clause that the contractor pay the host government a $2 million bonus when production for a certain period exceeds 10,000 BOPD. Another bonus may be required when production exceeds 20,000 BOPD. Such production bonuses stipulate a certain period such as a month or 120 days during which average production must exceed the specified level. Or perhaps a bonus would be required at production startup or when the 50 millionth barrel is produced. There are many variations on this theme. Payments are typically due within 30 days of the end of the month in which the obligation to make the payment occurs.

RIGHTS AND OBLIGATIONS OF NATIONAL OIL COMPANY

This section covers various issues that are not specifically covered in other sections of the contract. The rights and obligations of the national oil company may include:

- The right of access to the data acquired by the contractor
- The right to assist and expedite the contractor's execution of the work program
- The right to appoint its representatives with respect to the contract and the joint venture management team
- The right to remove at the contractor's expense any of the contractor's employees if the employee is incompetent, and/or unacceptable to the national oil company due to political or social behavior
- Use of contractor's equipment by the national oil company to the extent that it does not interfere with operations

RIGHTS AND OBLIGATIONS OF CONTRACTOR

The rights and obligations of the contractor may include:

- Fulfillment of all technical requirements
- Funds furnished for the performance of the work program
- The right to sell, assign, convey, or otherwise dispose of all or any part of its rights and interests under the contract to an affiliate or other parties with the prior written consent of the national oil company. It is hoped consent would not be unreasonably withheld. (see Ownership Transfer)
- Taxes

  Some contracts will specify what the tax rates will be. This is usually when tax rates are a negotiated item and the tax rates may be found in the Sales Proceeds Allocation clause (which goes by many names).

  When taxes are legislated, usually from a higher authority than the national oil company, the actual tax rate may not be specified and the contract will simply state that the contractor is responsible for all taxes.

- Submit weekly and monthly reports to the national oil company
- Give preference to goods and services available within the host nation
- Allow duly authorized representatives of the national oil company access to the area covered in the contract

VALUATION OF PETROLEUM

The valuation of crude petroleum exported from the host country is often defined at the international market price, or it is based on a predetermined "basket" of crudes. This price formula then provides the basis for determining taxes and the basis of cost recovery. This price may often be different than actual "realized" prices. Sometimes a committee consisting of contractor and national oil company personnel is established to monitor the international market price and maintain a realistic value of crude for transactions between the contractor and the national oil company.

The valuation issue is particularly critical if the contractor is paid in cash for services, as in a typical service agreement, or if the contractor has a crude purchase agreement with the national oil company.

PAYMENTS

There are often numerous obligations and corresponding payments between the host government agencies and the contractor. This clause stipulates the currency that will be used for payments between the contractor and the national oil company or other agencies. Most, but not all, contracts use U.S. dollars. Sometimes more than one currency is used, such as U.S. dollars and the local currency. When local currency is used, artificial exchange rates other than the floating market rate can cause problems.

RECOVERY OF OPERATING COSTS AND NET SALES PROCEEDS ALLOCATION

This section of the contract will explain the contractor's entitlement to recover all operating costs out of gross sales proceeds (unless the contract has a royalty). This section usually explains that if operating costs exceed the funds available for recovery, the remainder may be carried forward. It further outlines the allocation of revenues remaining after cost recovery. These remaining revenues are usually referred to as *profit oil* in the contract. Applicable taxes levied on the contractor's share of profit oil may be specified here, or the applicable body of tax law that applies will be specified.

EMPLOYMENT AND TRAINING OF LOCAL PERSONNEL

This section of the contract is usually an agreement by the contractor to employ as many qualified nationals and/or personnel from the national oil company staff as is possible, subject to competitive standards. Furthermore, this section outlines the timing of recruitment and training programs for the national oil company personnel.

This clause can appear to be toothless, in some cases requiring simply that *preference* to nationals be given. However, employing nationals is important, whether contractually required or not. In other cases there may be quotas that must be met. Some contracts will stipulate up to 85% employment of nationals.

TITLE TO EQUIPMENT

Most PSCs and service agreements specify that any equipment purchased by the contractor becomes the property of the national oil company as soon as it arrives in the country. The clause further stipulates that the contractor has the right to the use of the equipment in the petroleum operations. This clause will often further indicate that the cost of the equipment is recoverable. Under some contracts, title passes when the costs have been recovered under the cost recovery provisions of the contract.

Equipment that is rented or leased (and not purchased) by the contractor does not become the property of the national oil company and this clause grants contractor rights to freely import or export such equipment from the country.

OWNERSHIP TRANSFER

In some contracts the ownership transfer rights are covered under the Rights and Obligations of Contractor clause. Ownership transfer, also known as transfer of rights, or assignment, is an important aspect of contract negotiations. It deals with the contractor's ability to find partners. Assignments that involve affiliated companies are normally a simple formality requiring notification or an application to the government. Transfer of rights to a third party is a different matter. Most contracts will give the contractor right to assign "in whole or in part any rights, privileges, duties or obligations" to any third party acceptable to the national oil company or appropriate authority. The process under which a third party is introduced and considered for acceptance can be daunting. The determining factors are usually the financial integrity and to a lesser degree, technical capability, unless operatorship is being transferred. It is hoped that the transfer would be free of charges, fees, or related taxes, such as a capital gains tax.

Capital gains taxes or regulations on the transfer of rights can have a strong influence on the marketability of a concession for sale or farmout. Some contracts will cede to the national oil company a right of first refusal—the option of taking up the interests being transferred under the same terms offered. In China the national oil company has this right for any nonrelated party transactions.

The problem with this issue from the government perspective is that most governments resent companies making their licenses the focus of international commerce. They view the process often as one of promotion rather than exploration. The countries with more experience in these matters do not worry so much about this and understand the element of *partnering* which is such a big part of this industry.

The ideal wording from the contractor's point of view goes something like this:

> Contractor shall have the right to freely assign, transfer, or otherwise dispose of any part or all of its interests to any affiliated company and to sell, transfer, convey, or otherwise dispose of all or any part of its interests to parties other than affiliated companies or other contractor participants with the prior written consent of the Ministry, which consent shall not be unreasonably withheld or delayed. Assignments are exempt from all taxes, levies, and duties.

Other language that would be desirable from the point of view of the contractor but which is less common:

> Conditions for transfer must be approved by the Government; such approval shall not be unreasonably withheld. If within thirty (30) days after notification to the Government of the proposed transfer, the latter has

not made known its decision, the transfer will be deemed accepted. Transfers of any nature will be exempt of any taxes.

BOOKS AND ACCOUNTS

This clause stipulates who will be responsible for keeping the books and accounts and what rights the other parties have for auditing and inspecting the books. The accounting procedure to be used is normally specified and outlined in an appendix— Appendix C in this outline.

PROCUREMENT

This clause primarily deals with the government's desire to ensure that local participation is maximized for goods and services. Many of the details for regulating procurement may not be found in the PSC, though. The Malaysian contract does not specify what levels of expenditure will trigger the need for a tender or government approval, but these details are found in the joint operating agreement between the contractor and the national oil company.

Most budgets are submitted to the national oil company for approval and have set levels of expenditure for given projects or items that will require an authorization for expenditure (AFE). The procedures and limiting criteria for contract awards are usually specified according to whether or not they fall under exploration, development, or production operations.

For large capital items, the contracts usually require a competitive tender. The level of expenditure that requires tendering can have a big influence on the efficiency of operations. Generally, expenditures over $50,000 will be put out for bid.

For expenditures above $250,000 or so, a bid list may be required. The Vietnamese PSC requires international tendering for any work costing over $200,000. The government will publish a list of companies, and only those companies will be eligible for tendering for certain goods or services. If the contractor has a preferred supplier for a particular item, then it is important to get that supplier qualified for the bid list.

Between that level and, say, $1,000,000, the contractor would tender, but may have authority to choose without government approval. Above $1,000,000 or so, the government will usually have approval rights. The lower these thresholds are, the less efficient an operation can become.

JOINT OPERATING AGREEMENT

With government participation, an operating agreement will also be required. These operating agreements are similar to the JOAs used between industry working-interest partners. The PSC itself may indicate the requirement for a JOA. Here is an example from a recent Malaysian contract:

> [The participating companies and the Malaysian National Oil Company operating arm CARIGALI] . . . as Contractors shall, as a condition precedent to the signing of this Contract, enter into an agreement between themselves to provide for the procedures whereby they shall exercise their rights and fulfil their obligations as Contractors (such agreement hereinafter referred to as the "Joint Operating Agreement").

In the Malaysian contract, this article also stipulates the 15% entitlement (carried working-interest share) that CARIGALI has in the license.

FORCE MAJEURE

Force majeure is a French term meaning "an overpowering force or coercive power." Its application to a contract is to limit the liability of either party, contractor or national oil company, for nonperformance of contract obligations due to war, political disturbances, an act of state, riots, earthquakes, epidemics, or other major causes beyond human control. As mentioned previously, the term is sometimes defined in the Definition section of the contract.

Some negotiators would like to see bureaucratic delays included as a *force majeure* item, but in most cases that sort of language will not fly.

ARBITRATION

This clause describes the methods and rules by which disputes will be settled, should conflicts arise between the parties to the contract. More and more countries are willing to agree to international arbitration of disputes.

Typical arbitration clauses stipulate that contract disputes be arbitrated in a specific language and arbitration system. Often the arbitration clause stipulates that each side appoint an arbitrator, and these two arbitrators choose a third to make up a tribunal. The decision of the majority is usually final.

The inclusion of such nonjudicial dispute resolution mechanisms is becoming widespread and widely accepted. It can be done on an *ad hoc* basis where the arbitrators are determined by provisions of the contract, or there are numerous agencies which include:

- ICSID (International Center for the Settlement of Investment Disputes)
- ICC (International Chamber of Commerce)
- UNCITRAL (United Nations Committee on International Trade Law)
- AAA (American Arbitration Association)

In Equatorial Guinea the dispute resolution clause is an agreement to submit the dispute to ICSID for conciliation and, if unresolved after three months, for arbitration. A key question is, Is Equatorial Guinea a party to the ICSID convention? It may have signed the convention but not ratified it. Or it may not have even signed it. If the country has not signed and ratified the appropriate convention identified in the contract, then any dispute will drop down into the local courts. Without arbitration as the means for resolution of disputes, the local court system is usually the only other alternative.

In addition to the choice of method or institution and rules under which disputes will be arbitrated, the arbitration clause should state that the decision of the arbitrator or arbitral board shall be final and binding on both parties.

One advantage of arbitration is that it is private, unlike litigation, and often less expensive.

INSURANCE

Most contracts require that the contractor secure and maintain insurance with reputable international insurance companies satisfactory to the national oil company. Furthermore, the contractor is usually required to make sure that subcontractors are adequately insured.

TERMINATION

Contracts under many circumstances may be terminated by the contractor by giving 60–90 days written notice to the national oil company. This condition usually does not apply during the exploration phase of the contract if the contractor has not completely fulfilled the work obligations.

ENTIRE CONTRACT AND MODIFICATION

This clause usually states that the contract and exhibits constitute the entire agreement between contractor and national oil company, that any other representations not embodied in the contract will not apply. All prior negotiations, representations, letters of intent, or other agreements are deemed cancelled or merged within the contract.

Exhibit A: Description of Contract Area

This exhibit usually outlines the exact coordinates and physical boundaries of the contract area.

Exhibit B: Map of Contract Area

A map of the contract area is usually included and considered a standard part of the contract. Coordinates mentioned in Exhibit A are usually shown on the contract area map.

Exhibit C: Accounting Procedure

This exhibit usually specifies what currency will be used for the books and records, as well as what language. English is usually preferred, as are U.S. dollars. Such things as operating costs are defined. Items to be excluded from cost recovery are also stipulated.

Exhibit D: Management Procedure

This clause will often stipulate that the contractor is responsible for the execution of the work program. It may further outline the makeup of the management committee, format for meeting minutes, and responsibilities and functions of the management committee. This section may set forth subcommittees for procurement and accounting.

Other Contract Terms and Issues

Abandonment

The ownership issue always seems to surface with the subject of abandonment under a PSC. In theory, where ownership of the assets and facilities rests with the state, as it normally does with a PSC, there also rests the abandonment liability. This would be particularly true if the contract expired and operations passed to the national oil company before shut-down. The issue gets slightly abstract because abandonment is a normal procedure under standard oilfield operating practices, but there is no revenue stream available to recover costs of abandonment when that time comes. However, this is usually not such a complex matter. Abandonment costs can be estimated and amortized with the establishment of an interest-bearing reserve fund through cost recovery during the producing years. Properly structured, the abandonment cost would be anticipated, and the fund would mature at the appropriate date with sufficient funds to cover abandonment. This could be done with a straight-line method or a unit-of-production approach. This issue was not anticipated in many of the early PSCs.

ECONOMIC STABILITY, EQUILIBRIUM, OR INTANGIBILITY CLAUSES

There is much discussion about developing language for protection against additional taxes and levies that are enacted after contract signature. These clauses theoretically allow for an equalizing adjustment in the contract to neutralize the effect of any additional taxes that may be imposed. This would be an antidote for creeping expropriation. If the parliament increased income taxes, such a clause might require the national oil company to adjust the profit oil split in order to keep the contractor "economically whole." If taxes go up, the profit oil split is adjusted to keep the contractor take at the original contract level.

This would essentially give the national oil company more power than the taxing authority and would emasculate the taxing authority, as far as the petroleum industry was concerned. It is hard to imagine that a government body such as a parliament or a congress would effectively cede its authority to the national oil company.

Normally, contractors have no such economic stability guarantees, and it is unlikely that such clauses will flourish. There is an old saying: "What Parliament giveth Parliament may taketh away." This is an ancient and universal truth. Furthermore, why should petroleum companies receive special exemption from

171

raised taxes? This can be a dangerous political position for a company or an industry. It may be more appropriate to try to protect against legislation that discriminates against the petroleum industry.

The Ecuador Service Agreement has an equilibrium clause that allows for adjustment in the service fee formula in the event of tax level changes. The purpose of the adjustment is to effectively maintain the same economic terms that had been agreed upon in the contract. The Vietnamese PSC also has language to this effect. It addresses the possibility of a legal or administrative provision or amendment that might result in a modification to the terms of the contract. If there is an adverse effect on the contractor, Petrovietnam will take measures to guarantee a situation that is economically equivalent to that initially envisaged in the contract.

The Malaysian PSC also has a clause that addresses this issue.

> The terms of this Contract have been negotiated and agreed having due regard to the terms of the Petroleum (Income Tax) Act, 1967 as amended by the Petroleum (Income Tax) Amendment Act, 1976 and export duty in force on the Effective Date. If, at any time or from time to time there should be changes in the aforesaid or the introduction of any other legislation, regulations or order which imposes taxes and duties peculiar to the petroleum industry and not of general application, the effect of which would be to increase or decrease materially the liability of Contractors to pay petroleum income tax or such other taxes and duties, then the Parties may consult each other with a view of arriving at an equitable arrangement so as to take account of such changes.

The Chinese PSC has a broader clause:

> If a material change occurs to the Contractor's economic

benefits due to new laws, decrees, rules, and regulations or any amendment to applicable laws, decrees, and regulations made by the government, the Parties shall make necessary revisions to the Contract to maintain the Contractor's normal economic benefits.

The virtue of having such clauses is that perhaps it would be less likely that changes subsequent to contract signature would have an adverse effect. Or a change in taxes, for example, may be less likely. Furthermore, clauses such as these add some teeth to claims for compensation should changes occur.

RENEGOTIATION CLAUSES

Some contracts have provisions for review and renegotiation at various contract intervals. This kind of clause amounts to an "agreement to agree" at a later stage in the contract life. There are strong emotions on both sides of this issue. Many negotiators have doubts about the value of review clauses and believe that it is perhaps better to structure *flexible* terms that can respond to a variety of economic conditions. They prefer to nail down every contract item as precisely as possible.

Others believe that renegotiation clauses provide a means of reaching agreement on basic issues and leaving the more uncertain issues for a time when better information is available to decide. A good example is what is sometimes referred to as the *gas clause*. Some contracts will specify the sharing and tax arrangement specifically for a gas discovery. But in the international sector, gas is quite different than oil in many ways. Some of these gas clauses effectively state that in the event of a gas discovery, the parties to the contract will agree to negotiate further.

The Ghana PSC had a renegotiation clause that provided for a review of the contract terms at any time if significant changes occurred in the circumstances that had prevailed at the time of entry into the agreement. Some such clauses require periodic review.

The Chinese contract also has a gas clause. The thrust of

THE VIRTUE OF RENEGOTIATION CLAUSES

DAVID JOHNSTON © '94

You understand, sir, that Rudolf was not fully aware that his long-term position with you amounted to only one night a year!

Illustration 8–1

the clause is that:

Allocation of the natural gas shall be in conformity with general principles of allocation for the crude oil stipulated in the contract. However, the percentages of the allocation shall be adjusted by the parties through negotiations in light of actual conditions in the gas field so that the contractor shall be able to obtain a reasonable economic benefit.

It is nearly impossible to anticipate and account for everything that may occur during the life of an agreement. Very often a problem appears to be an obvious blunder—hindsight is crystal clear. This is portrayed in Illustration 8–1. Many readers may recall that Rudolph was recruited at the last moment with little time for negotiation.

CUSTOMS DUTIES EXEMPTIONS

Contractors are usually exempt from normal customs duties. One reason for this is that title to the equipment is transferred to the government anyway. Sometimes title will transfer as it clears customs.

Many companies would like to see the customs duties exemptions, as agreed in the contract, included in the nation's petroleum law. There are often conflicts of authority between the host government, the national oil company, and the customs and immigration authority. A PSC may clearly state that no delays or duties will be imposed at customs, but unless this is well understood and respected by the customs officials, then it may not matter what the contract says.

The language often takes the following form:

Contractor and its subcontractors are entitled to import or export, without duty and without undue delays, equipment, materials, vehicles, spare parts, foodstuffs, and other supplies that are necessary for operations.

EXCHANGE CONTROL

Some governments try to regulate currency exchanges and financial transactions that involve capital coming into or going out of the country. These exchange controls are usually administered by the central bank or a government board of control. The situation becomes a problem particularly where local currency is overvalued. The difference between official exchange rate in a country and the black market rate is often a direct measure of the magnitude of the problem.

GOVERNING LANGUAGE

The governing language of the contract is usually the national language of the country. Indonesia requires the contracts to be written in Indonesian and English, but the English text is official. In China the contract language is both English and Chinese with both versions having equal force. In Vietnam contracts are required in Vietnamese and French. This, of course, conjures up images of three translations.

9

ACCOUNTING PRINCIPLES

An understanding of basic accounting concepts and principles helps when confronting the numerous fiscal systems in the international petroleum industry. The basic issues that define generally accepted accounting principles (GAAP) are straightforward and usually based on common sense. Knowing the reasoning behind accounting conventions can provide a good framework for understanding some of the accounting aspects of various fiscal systems. Most fiscal systems worldwide follow GAAP as outlined here.

Furthermore, eligibility requirements for foreign tax credits usually revolves around whether or not the foreign tax payments conform to GAAP, particularly in the determination of taxable income.

The Financial Accounting Standards Board (FASB) in the United States, an independent, self-regulating organization, establishes standards known as GAAP for the accounting industry. The FASB outlines procedures and rules that define accepted accounting practices for financial reporting. The FASB publishes broad guidelines and detailed procedures for financial accounting practices.

ACCOUNTING CONCEPTS

Eleven basic accounting concepts provide the foundation of accounting theory. These principles are fundamental to the understanding of financial reporting and calculation of taxable income. They may not be perfect. In nearly every case where there is a weakness in a particular principle or accounting practice, it is easy to point out the problem, but not so easy to find a better solution.

ACCOUNTING CONCEPT

1. Money measurement
2. Entity
3. Going concern
4. Dual aspect
5. Accounting period
6. Materiality
7. Conservatism
8. Consistency
* 9. Realization
* 10. Matching
* 11. Cost

* These last three concepts are of prime importance in understanding financial reporting in the oil industry.

1. THE MONEY MEASUREMENT CONCEPT

Financial information is expressed in monetary terms. It would not be practical to record on the balance sheet the number of barrels or acres a company owns. Money is the common denominator, and quite often the U.S. dollar is the currency used for accounting and reporting.

2. THE ENTITY CONCEPT

Accounts are kept for business entities. The answer to any accounting issue must address the question, How does it affect the business? According to the entity concept, accountants are not concerned with persons who own or operate the business, but with the business itself.

3. THE GOING-CONCERN CONCEPT

Accounting assumes that a business will continue forever. If there is evidence, that the entity is going to liquidate, the accounting function might need to assess what the entity is worth to a potential buyer.

4. THE DUAL-ASPECT CONCEPT

The resources owned by a business or entity are called *assets*. The claims against these assets are called *equities*. The two types of equities are (1) liabilities that represent the claims of creditors and (2) the *owner's equity*. The total claims on the assets are equal to the assets.

Assets = Equities

Assets = Liabilities + Stockholders' Equity

This is why both sides of the balance sheet balance. The essential concept is that for every resource available to a company, somebody has a claim on it.

5. THE ACCOUNTING-PERIOD CONCEPT

Accounting practices are based on the need to report periodically the status of a business entity. The basic time period is the fiscal year (12 months). Many companies use interim reports, usually quarterly and even monthly reports.

6. THE MATERIALITY CONCEPT

Insignificant items do not require attention.

7. THE CONSERVATISM CONCEPT

This principle dictates that given a choice, an asset will be recorded at the lowest or most conservative value. As far as the income statement is concerned, this principle provides that potential losses be accounted for, and yet potential gains or profits are not registered until they are realized. The result, theoretically, is that financial statements will provide a *conservative* view of the business.

8. THE CONSISTENCY CONCEPT

The consistency concept stipulates that once an entity embarks on an accounting methodology, it must be consistent in its treatment of accounting issues unless it has good reason to do otherwise. At times, companies decide to make changes in accounting policies. These changes are explained in the footnotes.

9. THE REALIZATION CONCEPT

The realization principle dictates that revenue should be recognized only at the time a transaction is completed with a third party, or when the value is reasonably certain. One common concern centers on the applicability of the realization principle to the petroleum industry. Some feel this principle should not be applicable because the major asset of an oil and gas company is its reserves, and the value of a company's reserves is not directly reflected on the balance sheet.

Neither the balance sheet nor the income statement allow appropriate recognition for important oil and gas discoveries in the accounting period in which a discovery is made. When a company makes a major discovery, there is no mechanism for reporting the results from an accounting point of view. The impact on the income statement comes when the discovery begins to produce, yet economic value was *realized* the moment the discovery was made.

Another aspect of the realization concept is the *accrual method* of accounting for revenue and expenses and ultimately taxes.

Under this method, revenue is recorded as it is earned, or is said to have accrued, and does not necessarily correspond to the actual receipt of cash. This concept is important for the understanding of the Statement of Cash Flows and the concept of *cash flow*. For example, assume that a company sold 1,000 bbl of oil for $20/bbl, but had received only $17,000 by the end of the accounting period. From an accrual accounting point of view, revenues are recorded as $20,000.

Revenues	$20,000	
Beginning receivables	+1,000	
Cash flow potential	21,000	
Ending receivables	– 4,000	
Realized cash flow	17,000	= Sales less increase in receivables

The income statement would reflect $20,000 because the accrual method of accounting *realized* the income at the point of sale, not at the point of actual cash exchange. The balance sheet would show the $17,000 increase in cash as well as an increase in accounts receivable for the $3,000 not yet received. However, the actual cash received is $17,000.

10. THE MATCHING CONCEPT

The matching principle provides that revenues should be matched with the corresponding costs of producing such revenues. A serious accounting issue in the oil and gas industry deals with the matching principle because it is so difficult to match the costs of finding oil and gas with the revenues from production. Under GAAP the assets reported on the balance sheet consist of capitalized historical costs. Earnings are recognized as reserves are produced rather than when they are discovered or revised.

Two separate systems of accounting in the industry are based primarily on this issue as it pertains to the treatment of exploration costs. The two systems are called *Full Cost* and *Successful Efforts* accounting. While revenues are typically recognized when oil or gas is sold, the fundamental difference between the FC and SE accounting systems lies in how the corresponding costs of finding those reserves match those revenues.

11. THE COST CONCEPT

In accounting, an asset is recorded at its original cost. This cost is the basis for all subsequent accounting for the asset. The primary rationale behind the cost principle is that the value of an item may change with the passage of time, and determination of value is subjective. There is no subjectivity associated with the actual cost of an item.

Because of the cost principle, the accounting entry on the balance sheet for oil and gas assets usually has little to do with the actual value of the assets. For instance, if a company were to obtain a lease and then discover 1 MMBBL of oil, the accounting entry would not change because of the discovery. It would never reflect anything other than the associated costs less depreciation. Only the net tangible costs would represent all that oil.

In this example there is substantial appreciation in value that is ignored by GAAP. However, accountants do not ignore economic value completely. The cost principle provides that assets should be reflected on the balance sheet at cost, unless there has been a decline in their utility or economic value. Accountants do not mind an asset on the books at less than its true market value, but they are careful to keep accounting entries from exceeding economic value.

Costs associated with oil and gas exploration and production fall into four fundamental categories:

Lease Acquisition Costs. Costs associated with obtaining a lease or concession and rights to explore for and produce oil and gas.

Exploration Costs. Costs incurred in the exploration for oil and gas such as geological and geophysical costs (G&G), exploratory drilling, etc.

Development Costs. Costs associated with development of oil and gas reserves, such as drilling costs, storage and treatment facilities, etc.

Operating Costs. Costs required for lifting oil and gas to the surface, processing, transporting, etc.

Treatment of these costs is fairly straightforward. The one exception is the way that exploration costs are treated. This provides the basis for the two different accounting practices that are used in the industry.

SUCCESSFUL EFFORTS AND FULL COST ACCOUNTING

These two accounting methods, Full Cost (FC) and Successful Efforts (SE), can give very different results, on earnings, return on equity, and book value. Both systems follow as best they can the accounting principles of matching, realization, and cost, yet there is debate about which system is most appropriate. Primarily, the two systems differ on how capital costs associated with exploration drilling are treated. The main difference is that drilling costs of unsuccessful exploration wells are capitalized under FC accounting and expensed under the SE accounting system.

SUCCESSFUL EFFORTS ACCOUNTING

Before the 1950s virtually all oil companies used some form of Successful Efforts (SE) accounting. The rationale behind SE is that expenditures that provide no future economic benefit should be expensed at the time incurred. The SE approach will *expense* or *write-off* exploratory dry-hole costs in the accounting period in which they are incurred. This is similar to many other businesses that write-off business failures. Proponents of SE agree that only expenditures directly associated with the discovery of hydrocarbons should be capitalized. SE companies treat exploration expenditures like other companies would treat research and development. If a research results in a viable product, then capital expenditures are capitalized, otherwise the costs are expensed.

COST CENTERS

One of the main differences between the two systems results from the choice of size and use of cost centers. It is this difference that makes the largest financial impact. With SE, costs for a cost center can be held in suspense until it is determined if commercial quantities of oil or gas are present. With a well or lease as the cost center, costs are expensed if the well is dry and capitalized if it is a discovery. This can be quite a subjective decision. Sometimes the decision to drill a well may be held up because of the perceived impact on the financial statements during a specific accounting period should the well turn out to be unsuccessful.

FULL COST ACCOUNTING

The philosophy behind FC accounting is that costs of acquisition, exploration, and development are necessary for the production of oil and gas. This rationale acknowledges that dry holes are an inevitable part of the exploration effort. With FC the entire company can be a cost center, and all costs of exploring for oil and gas are capitalized. Companies with international operations typically treat each country as a separate cost center. By contrast, as mentioned earlier, successful-efforts companies may treat each well as a cost center.

Ceiling Test Limitation

FC accounting requires a write-down on the book value of oil and gas assets if it exceeds the SEC value of reserves. For this reason, FC companies use large cost centers. This is the *ceiling test* required by the SEC for the cost of oil and gas properties on the balance sheet. The recorded capitalized costs for producing oil and gas properties is limited to the net present value of the reserves discounted at 10%. This is the *SEC value of reserves* or *standardized measure*. If the SEC value of reserves falls below the capitalized costs on the balance sheet, a ceiling write-down occurs. For example, if a company had a book value for proved oil and gas properties of $100 million, and if the SEC value of these reserves was $130 million, there would be no write-down. The company would have a *cost ceiling cushion* of $30 million. In 1986, when oil prices dropped so dramatically, cushions disappeared, and many FC companies experienced substantial write-downs. The most important problem that this caused was that many companies suddenly found themselves in violation of covenants in their loan agreements.

Impairments and write-downs occur under SE accounting, too. It is usually not considered as great an issue because such a large part of exploration costs are expensed and not capitalized. But, in consistency with the conservatism principle, the carrying value of SE oil and gas properties is subject to write-downs if the economic value of a property is less than the recorded value. Periodic assessments are made to insure that the value of leases have not been impaired due to negative results of drilling or approaching expiration dates.

COMPARISON

With each system, lease bonus payments, related legal costs, and intangible drilling costs (IDCs) are capitalized. Capitalized costs within a cost center are usually amortized on the unit-of-production method (explained later in this chapter). Basic elements of the two systems are compared in Table 9–1.

Table 9–1

Comparison of Accounting Systems

	Successful Efforts Method	Full Cost Method
G & G costs	Exp	Cap
Exploratory dry hole	Exp	Cap
Lease acquisition costs	Cap	Cap
Successful exploratory well	Cap	Cap
Development dry hole	Cap	Cap
Successful development well	Cap	Cap
Operating costs	Exp	Exp
Which companies typically use each method	Major Oil Companies	Smaller Independent Companies
Size of cost center used	Small	Large
	Single Well, Lease, or Field	Company, Country, or Hemisphere
Comment	Favored by FASB	Approved by SEC

G & G = Geological and Geophysical
Exp = Expensed (written off in accounting period)
Cap = Capitalized (written off over a number of accounting periods)

DEPRECIATION, DEPLETION, & AMORTIZATION (DD&A)

Depreciation is a means of accounting for the recovery of the costs of a fixed asset by allocating the costs over the estimated useful life of the asset. When this concept is applied to a mineral resource such as oil and gas reserves, it is called *depletion*. The concept is called *amortization* when the allocation of costs is applied to intangible assets. The terms *depreciation, depletion,* and *amortization* (DD&A) are sometimes used interchangeably, or more often collectively as *DD&A*.

The combined DD&A for many companies can be quite significant. The per-unit values that are deducted for income tax purposes can range from $3 to more than $10/bbl.

The importance of DD&A is that these expenses are deducted from income for federal and provincial tax purposes or as an important component of cost recovery under a PSC. The depreciable life of an asset is usually determined by legislation to emulate the useful life of that asset. GAAP usually dictate the use of straight-line decline over the useful life of an asset. However, some forms of accelerated depreciation are allowed as well. Four methods of depreciation/amortization are summarized:

- Straight-line decline (SLD)
- Sum-of-the-year's digits (SYD) *
- Declining balance (DB) *
- Double declining balance (DDB) *
- Unit of production
     *Accelerated depreciation

STRAIGHT-LINE DECLINE

Under SLD depreciation the asset is amortized in equal installments over its useful life. Thus if a six-year life is used, the depreciation rate would be $1/6$ of the original value per year. Sometimes contract summaries identify the depreciation rate as a percentage, such as 25%. This nomenclature indicates that the asset is depreciated at a rate of 25% per year, the equivalent of a four-year straight-line decline.

SUM-OF-THE-YEAR'S DIGITS

SYD is based upon an inverted scale which is the ratio of the number of digits in a given year divided by the total of all years digits. For example, with a five-year SYD depreciation schedule, the five year's digits are added (5 + 4 + 3 + 2 + 1) to get 15. Depreciation in the first year then (because it is an inverted scale) is equal to $5/15$ of the asset value. Year 2 would be $4/15$ and so on.

SUM-OF-THE-YEAR'S DIGITS DEPRECIATION

$100,000 investment

Year	Fraction	Percent	Depreciation	Balance
1	5/15	33.333%	$33,333	$66,667
2	4/15	26.667%	26,667	40,000
3	3/15	20.000%	20,000	20,000
4	2/15	13.333%	13,333	6,667
5	1/15	6.667%	6,667	0
		100.000%	$100,000	

DECLINING BALANCE

With the DB method, depreciation is straight-line depreciation calculated for the remaining balance of the asset for each year. For example, a four-year DB would depreciate 25% of an asset in the first year. The following year 25% of the remaining balance (75%) would be depreciated (18.75%). The third year, 25% of the remaining balance (.25 × .5625) would be depreciated and so on.

DOUBLE DECLINING BALANCE

The DDB method calculates depreciation by doubling straight-line depreciation for the remaining balance of the asset for each year. For example, four-year DDB would depreciate 50% of an asset in the first year. The following year depreciation would be 50% of the balance, or 50% of 50% = 25%. The remaining balance would be 25%, so in the third year depreciation would be 50% of that.

FORMULA FOR UNIT-OF-PRODUCTION METHOD

Annual Depreciation $= (C - AD - S)\dfrac{P}{R}$

where:

C = Capital costs of equipment
AD = Accumulated depreciation
S = Salvage value
P = Barrels of oil produced during the year *
R = Recoverable reserves remaining at the
beginning of the tax year

* If there is both oil and gas production associated with the capital costs being depreciated, then the gas can be converted to oil on a thermal basis.

Table 9–2 compares four depreciation methods for an asset with a useful life for accounting purposes of seven years.

Table 9–2

DEPRECIATION SCHEDULES

	Input
Investment ($)	100
Year	7

	7 Year SLD		7 Year DB		7 Year DDB		7 Year SYD	
Year	Annual	Cum.	Annual	Cum.	Annual	Cum.	Annual	Cum.
1	14.29	14.29	14.29	14.29	28.57	28.57	25.00	25.00
2	14.29	28.57	12.24	26.53	20.41	48.98	21.43	46.43
3	14.29	42.86	10.50	37.03	14.58	63.56	17.86	64.29
4	14.29	57.14	9.00	46.02	10.41	73.97	14.29	78.57
5	14.29	71.43	7.71	53.73	7.44	81.41	10.71	89.29
6	14.29	85.71	6.61	60.34	5.31	86.72	7.14	96.43
7	14.29	100.00	5.67	66.01	3.79	90.51	3.57	100.00
8			33.99	100.00	9.49	100.00		
9								
10								

SLD = Straight-line decline
DB = Declining balance
DDB = Double declining balance
SYD = Sum-of-the-year's digits

DOUBLE
TAXATION

Most companies doing business overseas must deal with the effects of both foreign taxes and their home country tax treatment of foreign income. Under U.S. tax law, the entire income, regardless of the source, of a domestic corporation is subject to U.S. income taxes. In this context double taxation is a key issue of international operations.

Double taxation occurs when countries have different definitions of taxable income/or profits. It can also arise when a taxpayer or taxpaying entity resident (for tax purposes) in one country generates income in another country. It generally refers to situations where the same profit is taxed more than once in more than one country.

Because of the competitive nature of the global industry, there is little room for double taxation. Margins are thin enough and there are few, if any, projects or provinces with sufficient potential to allow an added burden of taxation. Without relief from double taxation, very little foreign exploration would be undertaken. If there were room for added home-country taxes on foreign upstream operations, then it would only be due to inefficient rent extraction on the part of the host country.

Relief from double taxation comes from double taxation treaties (bilateral or multilateral) or home-country regulations that allow relief. The most common remedy is where companies are allowed to offset home-country corporate taxes with tax credits from foreign-paid taxes. Since 1918 the United States has provided unilateral relief from double taxation by allowing foreign income taxes or taxes on profits to be used as a tax credit against U.S. corporate income tax liabilities. The policy outlined in Internal Revenue Code Section 901(a) allows a credit against foreign income taxes to alleviate the double taxation that would exist if foreign income earned were taxed both by the United States and the host government of the country in which the income was earned. Different forms of relief are summarized in Table 10–1.

TAX CONSIDERATIONS

In order to be eligible for creditability against U.S. taxes, a foreign levy must be a tax, and its predominant character must be that of a U.S. income tax.

A tax is a compulsory payment to foreign country pursuant to its authority to levy taxes. Consumption taxes, penalties, and customs duties would not qualify as a tax in this sense. According to the U.S. Internal Revenue Code Section 901, creditable foreign taxes include "income, war profits, and excess profits taxes paid or accrued . . . to any foreign country or any possession of the United States."

PREDOMINANT CHARACTER

U.S. regulations dictate that in order to qualify, the foreign tax must be an income tax; that is, its *predominant character* must be that of a U.S. income tax predominantly consistent with U.S. income taxation principles. The purpose of the U.S. foreign tax credit system is to alleviate double taxation on income taxes.

Therefore, nonincome-type taxes such as property taxes and consumption taxes are not creditable. There are two primary tests for the predominant character requirement. The first is referred to as the *net gain* requirement.

Under the normal circumstances under which the tax applies, the foreign tax must be likely to reach a net gain. There are three main requirements.

1. REALIZATION REQUIREMENT

The foreign tax law must generally conform to the realization principle of accounting. The tax is imposed either during or after the occurrence of a taxable event—not before. If income or gain is recognized and taxed before when it would have been realized and recognized under generally accepted U.S. accounting principles, then the realization test is not met.

2. GROSS RECEIPTS REQUIREMENT

Calculation of the tax must begin with gross receipts or be predominantly based upon gross receipts.

3. NET INCOME REQUIREMENT

The significant costs incurred in operations reaping the gross receipts must be allowed as deductions in calculating taxable income. The timing of the recovery of costs must allow for all substantial costs (or more) to be recovered.

Second, if a foreign tax is contingent upon creditability of that tax against the taxpayer's home-country income taxes, it is referred to as a *soak-up tax,* and U.S. regulations will not allow it to be credited.

There are other limitations—for example, foreign royalties will normally not qualify. Royalties and bonuses are considered to be payment for *specific economic benefits* that are not ordinarily available on substantially the same terms to all taxpayers in a given country. This situation is typical for the petroleum industry where the specific benefit is usually the right to extract hydrocarbons. Taxpayers that receive a specific economic benefit and are also subject to pay taxes are referred to as *dual-capacity taxpayers*, because they both receive a benefit and pay taxes.

If a foreign government owns mineral resources in which the taxpayer has an interest, a foreign tax will not be recognized as a tax for U.S. federal income tax purposes, unless that government also requires payment of an appropriate royalty or other consideration for the payment that is commensurate with the value of the concession or license—the specific economic benefit. The foreign income tax must be calculated separately and independently of the amount of the royalty and of any other taxes, payments, or charges imposed by the foreign government. If the foreign tax does not impose a royalty, the IRS allows dual-capacity taxpayers to separate the levy into its separate components—creditable tax payment and royalty payments or the equivalent. The dual-capacity taxpayer must also prove the creditability of the tax portion or the *qualifying amount* of the levy. There are two methods for separating a foreign levy into its distinct creditable and noncreditable elements—the *facts and circumstances method*, and the *safe harbor method*.

In the facts and circumstances method, the dual-capacity taxpayer establishes which portion of the foreign levy is not paid for the specific economic benefit.

The safe harbor method is based upon an IRS Code formula:

SAFE HARBOR FORMULA

Creditable portion of levy $= (A - B - C) \times D/1 - D$

where:

A = the amount of gross receipts

B = the amount of costs and expenses

C = the total amount foreign levy paid by the dual-capacity taxpayer

D = general tax rate

For example, assume that Teton Exploration (a dual-capacity taxpayer) is subject to a 50% income tax in Country X where the general tax rate is 35%. The Country X tax on Teton income is a separate tax than the Country X general income tax structure. The safe harbor formula is used to calculate the creditable portion of the levy and to separate the payment for the specific economic benefit. Assume that Teton has $20 million in gross receipts and $12 million in costs and expenses. The resulting reported net income of $8 million is taxed at a rate of 50%.

Calculation of the eligible foreign tax credit (in millions of dollars) follows:

$$FTC = (20 - 12 - 4) \times \frac{.35}{1 - .35}$$

$$FTC = \$4 \times .53846$$

$$FTC = \$2,153,840$$

The noncreditable payment for the specific economic benefit is $1,846,160—the difference between the foreign tax credit and the actual payment ($4 million).

$$\$4,000,000 \quad \text{Actual tax paid}$$
$$- 2,153,840 \quad \text{Creditable portion}$$

$$\$1,846,160 \quad \text{Payment for specific economic benefit, noncreditable}$$

The regulations in this area are quite lengthy.

Taxes in Lieu of Income Taxes

Some foreign taxes that do not meet the net gain require-ments may still qualify for creditability. A foreign tax that is imposed "in lieu of a tax on income" may be treated as an income tax. A foreign tax will qualify as an *in-lieu-of* tax if it is a tax, it is not a soak-up tax, and if a substitution requirement is met. Under the substitution requirement, a tax is an in-lieu-of tax if it oper-ates as a substitute to and not in addition to an income tax. It need not be a complete substitute.

Relief from the burden of double taxation comes in three basic forms: exemptions, tax credits, and deductions.

EXEMPTIONS. Under the exemption system, profit is taxed in one country and is exempt from tax in the others.

TAX CREDITS. A common system is where income taxes paid in a source country are creditable against income taxes in the country of residence. This system means that the effective tax is the high-er of the two country's tax rates.

DEDUCTION. Under this system, taxes paid in the source country are deductible against income taxes in the country of residence. For example, in Papua New Guinea the Basic Petroleum Tax is creditable against U.S. income taxes but the Additional Petroleum Tax (which is essentially the basis for classifying the PNG contract as a ROR contract) is not creditable. However, the APT is deductible when calculating U.S. income taxes.

The Indonesian 35% income tax is creditable against U.S. income taxes. The Indonesian production share and the 20% withholding tax do not qualify. The 35% tax is consistent with the general IRS guidelines for tax creditability under Section 901:

1. The amount of tax is calculated separately and independently of the amount of royalty (if any) and other taxes or charges imposed.

Table 10–1

Double Taxation Relief

	System			
	No Relief	Tax Deduction	U.S. Tax Credit	Tax Exemption
Profits in Country X	$100	$100	$100	$100
Country X Income Tax (26%)	26	26	26	26
After-tax Income in Country X	74	74	74	74
Home-country Tax Basis	100	74	100	0
Home-country Tax (34%)	34	25	8	0
After-tax Income	40	49	66	74
Total Taxes Paid	60	51	34	26
Effective Total Tax Rate	60%	51%	34%	26%

2. The income tax is imposed on the receipt of income determined on the basis of arm's length transactions and in accordance with U.S. income taxation principles.

3. The taxpayer's liability cannot be discharged from property owned by the foreign government.

4. The tax liability is computed on the basis of the taxpayer's entire extractive operations within the country.

5. Reasonable limitations on the recovery of capital expenditures is allowed, but the taxpayer must be able to deduct without limitation the significant expenses incurred.

In some production sharing contracts, there is not a direct royalty or tax payment by the contractor, and the language of the contract effectively states that royalties and taxes are paid by the national oil company out of its share of profit oil. The Egyptian PSC is a good example of this. The income tax liabilities of the contractors in Egypt were paid by EGPC out if its share of profit oil. The IRS allowed companies to credit the Egyptian income tax liability against U.S. tax liabilities in spite of the arrangement.

The 1971 and 1978 Peruvian model contracts with no direct cost recovery provisions created a problem. These contracts simply gave the contractor a share of production. This kind of arrangement was even further removed from the general IRS guidelines than the Egyptian contract. In order to achieve eligibility, Belco Petroleum and Occidental Petroleum negotiated with the Peruvian government for a contract revision which would have resulted in an option under which the companies could pay a 40% tax on gross income or a 68.5% net income tax paid directly to the government. The proposed agreement included a royalty to be paid to the national oil company PetroPeru. However, in 1980 the Peruvian government decided against the gross income tax and passed legislation for a tax on net income.

LIMITATION ON AVAILABLE TAX CREDIT

If income from a foreign operation is taxed at a rate that is higher than the U.S. tax rate, then the total creditable taxes could exceed the U.S. tax on that foreign income. However, the United States limits the amount of credit on foreign income to the amount of U.S. tax on that income. The U.S. approach to the taxation of income from foreign operations is based on the principle that foreign operations should face a total tax rate at least equal to the U.S. tax rate.

BRANCHES VS. SUBSIDIARIES

U.S. companies operating overseas in the past have normally made use of a *branch* or a special-purpose, Delaware-incorporated *subsidiary*.

BRANCH

When a foreign operation is a branch of the home-country parent organization, no intercompany dividend is involved, and normally withholding taxes are not levied. For example, a company operating as a branch in New Zealand does not pay the 15% withholding tax, but the income tax rate is higher for a branch at 38% compared to the normal rate of 33% for domestic corporations and foreign-controlled subsidiaries. Some countries treat payments from branches to parents as dividends for withholding tax purposes, but this is very rare. Prior to 1984 in the United States, the rule of thumb was, "You always *drilled* in branches." This was because losses from a branch's operations could be offset against other domestic income. At a later date, when the branch began to throw off profits, it could be incorporated in the foreign jurisdiction to allow the potential deferral of taxation on the profits. Companies can no longer do this because prior branch foreign losses have to be recaptured when a foreign branch is incorporated. Now the default option is usually to drill through a *controlled foreign corporation* (CFC). If the drilling is unsuccessful, and assuming the corporate structure is properly set up, the stock of the CFC can be taken as a loss under IRC Section 165(g)(3). If the drilling is successful, it may be possible to defer paying any current U.S. tax on the foreign earnings—depending on the classification of the income under Subpart F (which will be discussed later).

SUBSIDIARY CORPORATION

If the foreign operation is structured through a subsidiary company, a controlled foreign corporation (CFC), it will pay income taxes on profits generated in the foreign country and also pay dividend withholding taxes when they are repatriated to the parent company. These withholding taxes are usually 15%–20%. The foreign tax credit availability is limited to shareholders with 10% or more of the voting stock of the CFC. To qualify as a CFC in the United States for eligibility for tax credits, U.S. shareholders must own more than 50% of the voting stock or value of the company.

A U.S. shareholder is a U.S. person who owns 10% or more of the voting power of the foreign corporation. Thus there is some planning room here to allocate nonvoting shares disproportionately to either create a CFC (i.e., to allow creditability of taxes) or not do so (i.e., where the taxpayer desires to let profits build up untaxed offshore).

Subsidiary companies are normally taxed when profits are remitted to the parent, as opposed to branch profits, which are taxed as they accrue.

Tax credits from international operations can come directly through a branch where the taxes are paid directly or indirectly through a subsidiary where the taxes are deemed to have been paid. There are some other differences between direct and indirect credits. Direct credits are operative in the year the taxes are paid, whereas the indirect credit, except when Subpart F is applicable, is triggered when income is distributed. The most common form of indirect tax credit is when a domestic company receives a dividend from a foreign subsidiary. The direct credit can be claimed by either an individual or a corporation but the indirect, or *deemed-paid*, credit can only be claimed by a corporation. The parent company must own at least 10% of the voting stock of the subsidiary. The amount deemed paid is based upon a proration of the actual amount paid and the percentage interest held by the parent.

The actual computation of the deemed-paid credit can be formidable in practice. The concept is to "gross up" the dividend and attribute the foreign tax paid with respect to that dividend, taking into account the foreign corporation's earnings and profit layers and the domestic parent's ownership percentage.

Significantly, the code allows a deemed-paid credit for a second-tier and third-tier subsidiary, provided the domestic parent has at least a 5% indirect interest in these entities. Thus, if a third-tier subsidiary produces income which generates a foreign tax credit and then pays a dividend which eventually goes to the domestic parent, the domestic parent may take a deemed-paid foreign tax credit.

Registration of a company in a foreign country usually requires the following:

- Appointing a local agent
- Maintaining a registered office in the country
- Lodging a certified copy of the certificate of incorporation and other documents (financials, annual reports, etc.) verifying the company's existence and validity with the registrar of companies or equivalent authority
- Submitting a certified list of directors

This topic would not be complete without a discussion of Subpart F. Subpart F seeks to treat certain kinds of income earned by CFCs as currently taxable to the U.S. shareholders in a manner similar to, but certainly not just like, an S Corporation, regardless of whether the profits have been repatriated to the U.S. shareholders. In a similar vein, a foreign tax credit may be taken on a deemed-paid basis for any income recognized by U. S. shareholders through the application of Subpart F.

Subpart F is particularly important in the oil and gas industry, which is given special treatment. A special limitation is imposed on the amount of foreign taxes taken into account on the income from Foreign Oil and Gas Extraction Income (FOGEI) and Foreign Oil Related Income (FORI). The purpose of these rules is to prevent allocation of FTCs out of the FOGEI category (which is usually taxed at a high rate) to FORI, or some other classification, which may be taxed at a lower rate, thereby enhancing the overall use of FTCs.

Foreign tax credit usage may be restricted under certain rules that require the recapture of overall foreign losses from a branch's operations. These rules work by re-sourcing foreign income as domestic income to the extent prior foreign losses had reduced domestic taxable income. Although only 50% of the otherwise foreign source income is recharacterized in a taxable year—thereby limiting the rule's effect to some degree—special rules may apply where a foreign property is disposed of that could re-source more than 50% of the otherwise foreign source income.

Certain rules called *dual consolidated loss provisions* apply to both domestic corporations and to separate units of the domestic subsidiary (i.e., branches). These rules prevent a U.S. company from "double dipping" by deducting losses from its worldwide operations under the normal U.S. tax rules and again in a foreign jurisdiction where the foreign jurisdiction allows losses to be taken on other than a source basis.

As mentioned before, a branch's losses must be recaptured when the branch is incorporated in a foreign jurisdiction. The amount recaptured is reduced by any overall foreign losses previously recaptured as described earlier. Other rules may apply which could reduce the amount recaptured. The important point to keep in mind is that these rules only apply if there are foreign losses attributable to the branch being incorporated. If there are no foreign losses, it may be possible to incorporate the branch tax-free if it constitutes an active trade or business. Again the IRS has devised special rules for the oil and gas industry, that apply here to determine whether or not an oil and gas working interest constitutes an active trade or business.

COMMENTARY

The international petroleum industry has entered an era of intense global competition. More countries than ever are open for business and are aggressively seeking capital. In 1993 there were official licensing rounds in 40 countries. Other nations were actively negotiating with oil companies. Republics of the former Soviet Union dominate the scene. This has created a heightened competitive awareness among countries all over the world. They now actively compete with one another. It is no longer a passive market. Countries must be proactive, and their marketing efforts have to be consistent with their culture, geology, and fiscal terms.

Whole regions such as Southeast Asia, Latin America, and West Africa are being evaluated on the basis of prospectivity, fiscal terms, and general business environment. Ultimately, countries within certain geographic regions may need to cooperate. Individual countries may not be able to stand alone.

Petroleum fiscal designs are evolving. Clear trends have been established. Most governments are relying on production sharing contracts, lower royalties, and fewer signature bonuses.

Exploration risks are high enough without bonuses. Competition and enlightenment have fostered these trends.

Governments are more flexible in awarding blocks and licenses. It is difficult for countries to design fiscal terms appropriate for every geological province. Countries can let industry help determine what the market can bear by negotiating terms.

Independent oil companies are going overseas as never before, entering the domain of the major oil companies. Exploration target thresholds are three to four times higher for a major than an independent. This inefficiency has added an important dynamic to the international business. Risk capital is too critical, and dinosaurs will not survive in the exploration business. Fewer and fewer regions are capable of yielding the major oil company exploration thresholds of more than 400 MMBBLS of oil.

As the international sector expands, oil provinces in the United States have entered a stage of supermaturity. No other region comes close. Figure 11–1 illustrates the worldwide picture as of January 1993. Onshore in the lower 48 states the average oil well produces less than 10 BOPD. By international exploration standards, the onshore United States is dead.

The future for the upstream oil industry lies outside the continental United States. In regions such as Latin America, Africa, Asia, and the former Soviet Union, opportunities will proliferate as these regions mature. Competition will help forge realistic and balanced fiscal terms as well as improved business climates. It is an exciting future with many challenges. Good luck!

– Daniel Johnston

Regional Reserves Distribution

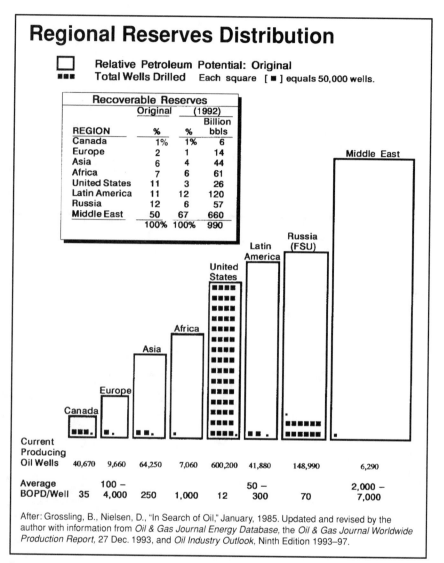

□ Relative Petroleum Potential: Original
■■■ Total Wells Drilled   Each square [■] equals 50,000 wells.

Recoverable Reserves			
	Original	(1992)	Billion
REGION	%	%	bbls
Canada	1%	1%	6
Europe	2	1	14
Asia	6	4	44
Africa	7	6	61
United States	11	3	26
Latin America	11	12	120
Russia	12	6	57
Middle East	50	67	660
	100%	100%	990

	Canada	Europe	Asia	Africa	United States	Latin America	Russia (FSU)	Middle East
Current Producing Oil Wells	40,670	9,660	64,250	7,060	600,200	41,880	148,990	6,290
Average BOPD/Well	35	100 – 4,000	250	1,000	12	50 – 300	70	2,000 – 7,000

After: Grossling, B., Nielsen, D., "In Search of Oil," January, 1985. Updated and revised by the author with information from *Oil & Gas Journal Energy Database*, the *Oil & Gas Journal Worldwide Production Report*, 27 Dec. 1993, and *Oil Industry Outlook*, Ninth Edition 1993–97.

Figure 11–1 Regional reserves distribution

APPENDICES

APPENDIX A
SAMPLE FISCAL SYSTEMS

The following examples outline basic elements of the fiscal systems from selected countries. The terms outlined here are nonproprietary, public domain information gathered from a variety of sources. Where appropriate, the source has been given. In most cases the terms are general. There may be some variation within a given country, but the terms outlined should be representative. It is believed that the summaries are complete in regard to the key royalty, tax, and production-sharing elements. However, where the information is not available or is ambiguous regarding such things as ringfencing or exploration obligations, the countries were nevertheless included.

COUNTRY FISCAL TERM SUMMARIES

1	Abu Dhabi	22	Ivory Coast
2	Albania		(Côte d'Ivoire)
3	Algeria	23	Korea
4	Angola	24	Malaysia
5	Argentina	25	Malta
6	Bangladesh	26	Morocco
7	Bolivia	27	Myanmar
8	Brunei		Myanmar Contracts
9	Cameroon		(IOR, RSF, Offshore)
10	China Offshore	28	New Zealand
11	China Onshore	29	Nigeria
12	Colombia	30	Norway
13	Congo	31	Pakistan
14	Egypt	32	Papua New Guinea
15	Equatorial Guinea	33	Philippines
16	France	34	Spain
17	Gabon	35	Syria
18	Ghana	36	Thailand
19	India	37	Timor Gap
20	Indonesia	38	Tunisia
	(2nd Generation Pre-1984)	39	Turkmenistan
	Indonesia	40	United Kingdom
	(4th Generation Post-1988–89)		History
21	Ireland	41	Uzbekistan
		42	Vietnam
		43	Yemen (North)

ABU DHABI
Early Concessions

Area	1 MM + Acres
Duration Exploration Production	2 years 33 years
Relinquishment	?
Exploration Obligations	Seismic Survey + 1 or more wells aggregate depth of 30,000 ft
Royalty	Sliding Scale: Up to 100,000 BOPD 12.5% 100,001–200,000 16% 200,001 + 20%
Bonuses	Commercial Discovery $5 MM 50,000 BOPD $3 MM 100,000 BOPD $6 MM 200,000 BOPD $6 MM
Cost Recovery	100% No Limit
Depreciation	5-year straight line
Taxation (Income Tax)	Sliding Scale, BOPD Up to 100,000 55% 100,001–200,000 65% 200,001 + 85%
Ringfencing	Each license separate
Domestic Market Obligation	Nil
State Participation	Some

Later modern contracts in Abu Dhabi followed the OPEC model:
20% Royalty
85% Tax

ALBANIA
Circa 1991

Area	No restriction, Designated blocks
Duration Exploration Production	2 + 3 + 1.5 years with discovery 4-year development period + 20 years
Relinquishment	25% or 100% of no discovery
Exploration Obligations	Seismic work programs offshore Start within 3 months
Royalty	Nil
Signature Bonus	Nil
Production Bonus	25,000 BOPD: $1.0 million 50,000 BOPD: $1.0 million
Cost Recovery	45% limit
Depreciation	20%/year for 5 years
Profit Oil Split (In favor of government)	Approximately 60%/40%
Profit Gas Split	Negotiable
Taxation	50% income tax
Domestic Market Obligation	Nil
State Participation	Nil

ALGERIA
Royalty/Tax (Law N° 86-14 of Aug 19, 1986)
Partnership Contracts & PSCs (Law N° 91-21 of Dec 4, 1991)

Area

Duration
  Exploration   4 years + 2-year extension
  Production   12 years from date of exploration permit

Relinquishment

Exploration Obligations

Royalty   Zone N   20%
                          Area A   16.25%
                          Area B   12.25%
Under certain conditions the 20% royalty may be reduced in Areas A & B

Bonuses   None for exploration permits
                       Negotiated for existing fields

Depreciation   G&G Costs and dry holes 100%
                       Producing Wells 25%/year
                       Buildings & Facilities 20%/year
                       Pipelines 10%/year

Taxation   Income Tax   Zone N   85%*
                          Area A   75%*
                          Area B   65%*
Under certain conditions the 85% tax rate may be reduced in Areas A & B.
The 1980 Amoco contract had 55% tax rate.
* Only valid with partnership contracts

Ringfencing

Domestic Market Obligation   For gas, negotiated formula depends
                                 on size of discovery

State Participation   51% at discovery to Sonatrach (NOC)
                       Carry through exploration + two
                       extension wells

Other

ANGOLA
1989 Model PSC

Area	Designated blocks
Duration Exploration Production	5 years (3 + 2 one-year extensions) 20 years
Relinquishment	?
Exploration Obligations	
Royalty	Nil
Signature Bonus	Yes, & production & education bonuses
Cost Recovery	50% limit Old contracts had 61% uplift for development costs recovered over 6 years.
Depreciation	5-year straight line

Profit Oil Split (Two examples)	Cumulative Production, MMBBLS	Split, % Gvt/ Company	Cumulative Production, MMBBLS	Split, % Gvt/ Company
	Up to 25	60/40	Up to 25	45/55
	25–50	70/30	25–75	70/30
	50–100	80/20	75–175	80/20
	Over 100	90/10	Over 175	90/10
on field by field basis				

Taxation	50% income tax
Ringfencing	Each license separate usually
Domestic Market Obligation	Nil
Investment Uplift	40% of tangible capital costs
Other	Old contracts had $20/bbl (1987 price cap). Gvt took excess above price cap escalated by U.N. factor for manufactured goods exports from developed nations. 1992 price cap = $35/bbl
State Participation	Yes, back-in option for development up to 51%—less in deepwater

ARGENTINA
Royalty/Tax (1990)

Area	Designated blocks
Duration Exploration Delineation Production	 3 years with two 2-year extensions 1 following discovery 20 years
Relinquishment	
Exploration Obligations	
Royalty	12% 5% marginal fields
Bonuses	
Depreciation	
Taxation (Profit Tax)	30% 1% sales tax 1% assets tax
Ringfencing	
Domestic Market Obligation	Yes
State Participation	15%–50% gvt option (old contracts)

BANGLADESH
(1989)

Area	Designated blocks
Duration Exploration Production	 4 years + two 2-year extensions 15 years from date of 1st sale
Relinquishment	After: 4 years 25% 6 years 50% 8 years 100% with no discovery
Exploration Obligations	1 well minimum
Royalty	Nil
Signature Bonus	No
Production Bonus	5,000 BOPD: U.S. $0.5 million 10,000 BOPD: 1.0 million 15,000 BOPD: 1.5 million 20,000 BOPD: 2.0 million
Cost Recovery	Sliding Scale, BOPD Up to 5,000 40% Limit 5,001–10,000 35% 10,001 + 30%
Depreciation	10% year straight line
Profit Oil Split (in favor of government)	Production, BOPD Split, % Up to 5,000 70/30 5,000–10,000 75/25 10,001–25,000 80/20 25,001–50,000 85/15 50,001 + 90/10
Profit Gas Split	Negotiated
Taxation	Nil, profit split is effectively after-tax split
Domestic Market Obligation	Pro rata up to 25% at market price
State Participation	Nil

BOLIVIA
Operation, Association, and Service Contracts

Area	Not to exceed 1 MM hectares in traditional areas Not to exceed 1.5 MM hectares in nontraditional areas 20,000 hectare lots
Duration Exploration Production	Maximum Term 30 years 4 years + 2 if gas is discovered 24–26 years depending on extension
Relinquishment	Exploitation area following exploration phase may not exceed 3 lots (60,000 hectares)
Exploration Obligations	
Bonuses	
Depreciation	
Royalties	19% national tax based on gross production 11% departmental participation royalty 1% national compensatory royalty
Taxation	40% net profits tax
Ringfencing	
Domestic Market Obligation	
State Participation	
Other	

BRUNEI
Royalty/Tax
From PetroAsian Business Report, March 1994

Area	No limitations
Duration Exploration Production	8 years onshore, 17 offshore Total of 38 years onshore, 40 offshore 30-year extensions may be available
Relinquishment	75% after exploration period if discovery is made
Exploration Obligations	B$4,000/km$_2$ seismic or B$12 million, whichever is greater, in the first 4 years in the next 4 years at least B$8,000/km$_2$
Royalty	12.5% onshore 10% for fields 3–10 nautical miles offshore 8% for fields more than 10 miles offshore
Bonuses **Rentals**	May be required—negotiable B$15/km$_2$ during first 4 years B$45/km$_2$ thereafter
Cost Recovery Limit	None
Depreciation	?
Taxation	55% petroleum income tax
Ringfencing	
Domestic Market Obligation	
State Participation	State has option to take up to 50%
Other	

CAMEROON
(1990)

Area	Designated blocks

Duration
Exploration	4 years + three 4-year renewals
Production	25 years

Relinquishment

Exploration Obligations

Royalty

Sliding Scale:

Up to 50,000 tons/yr	(1,000 BOPD)	2%
50,001–400,000 tons/yr	(7,700 BOPD)	6%
400,001–700,000 tons/yr	(13,400 BOPD)	9%
700,001–1,000,000 tons/yr	(19,000 BOPD)	11%
Over 1,000,001 tons/yr	12.5%	

Gas less than 300 billion m³/yr (29 MMCFD)	1%
more than 300 billion m³/yr	5%

Signature Bonus

Depreciation

Profit Oil Split

(In favor of government)

Cumulative Production	Split %
Up to 15 MM tons (100 MMBBLS)	60/40
15–30 MM tons (200 MMBBLS)	65/35
Over 30 MM tons (200+ MMBBLS)	70/30

Taxation	57.5% for petroleum companies

Ringfencing

Domestic Market Obligation

State Participation	50% state reimburses contractor for exploration costs

CHINA Offshore PSC
From *PetroMin Magazine*, Dec. 1993

Area	?

Duration	30 years
Exploration	7 years
Production	15 years + extensions with approval

Relinquishment	?

Exploration Obligations	?

Royalty

Oil, BOPD		Gas, MMCFD	
Up to 20,000	0%	Up to 195	0%
20,001–30,000	4%	195–338	1%
30,001–40,000	6%	338–484	2%
40,001–60,000	8%	484 +	3%
60,001–80,000	10%		
80,001 +	12.5%		

(BOPD converted from tons/year at 7:1) : (MMCFD converted from MM m₃/year at 35.3:1)

Pseudoroyalty	5% consolidated industrial and
Based on Gross Revenues	commercial tax

Profit Oil Split

(Negotiable) Production Rate, BOPD	Example Split, % Gvt/Contractor
Up to 10,000	10/90*
10,000–20,000	20/80
20,000–40,000	30/70
40,000–60,000	40/60
60,000–100,000	50/50
Over 100,000	60/40

*Some contracts start at 95% and slide to 45%.

Bonuses	?

Cost Recovery Limit	50%–62.5%
With 9% interest cost recovery on development costs	

Depreciation	6-year SLD

Taxation	30% income tax
	(15% in Hainan Province)
	3% local income tax
	10% surtax
Contractors must also pay vehicle and vessel usage, license tax, and individual income tax.	

Ringfencing	Yes, for cost recovery but not for income tax purposes

Domestic Market Obligation	No

State Participation	Up to 51% upon commercial production

CHINA Onshore PSC
From *PetroMin Magazine*, Dec. 1993

Area	?

Duration	30 years maximum
Exploration	8 years (3 + 2 + 3)
Production	15 years + extensions with approval

Relinquishment	?
Exploration Obligations	?
Bonuses	?

Royalty

Oil, BOPD		Gas, MMCFD	
Up to 1,000	0%	Up to 10	0%
1,001–2,000	1%	10–20	1%
2,001–3,000	2%	20–30	2%
3,001–4,000	3%	30–40	3%
4,001–6,000	4%	40–60	4%
6,001–10,000	6%	60–100	6%
10,001–15,000	8%	100–150	8%
15,001–20,000	10%	150–200	10%
20,001 +	12.5%	200 +	12.5%

(BOPD converted from tons/year at 7:1)   (MMCFD converted from MM m₃/year at 35.3:1)

Pseudoroyalty	5% CIC Tax based on gross revenues
Based on Gross Revenues	Commercial Tax

Profit Oil Split (Negotiable)

Production Rate, BOPD	Example Split, % Gvt/Contractor
Up to 10,000	10/90
10,000–20,000	20/80
20,000–40,000	30/70
40,000–60,000	40/60
60,000–100,000	50/50
Over 100,000	60/40

Cost Recovery	60% onshore

With 9% interest cost recovery on development costs

Depreciation	6-year SLD

Taxation	30% income tax (15% in Hainan Province)
	3% local income tax
	10% surtax

Contractors must also pay vehicle and vessel usage, license tax, and individual income tax.

Ringfencing	Yes

Domestic Market Obligation	No

State Participation	Up to 51% upon commercial production

COLOMBIA
Mid 1980s Association Contract, pre-1994

Area	Various blocks
Duration	28 years including exploration period Up to 10 years for exploration
Relinquishment	50% at end of year 6 25% more at end of year 8
Remainder is dropped at end of year 10 except commercial field and 5 km band.	
Exploration Obligations	Yes, negotiated
Royalty 1990 War Tax	20% 600–900 pesos/bbl fluctuating but based on $1.00/bbl for 1st 6 years of production
Bonuses	
Cost Recovery	100% No limit
The term "cost recovery" is not used, but contractor collects all of gross revenues except royalty (i.e., 80%) until payout.	
Depreciation	

Taxation	30% income tax
	25% surcharge from 1993–97
	   effective rate 44%
	   (expected reduction to 36%)
	   Remittance Tax 12%–15%

Ringfencing	No
Domestic Market Obligation	Receives FMV 75% in U.S. $
State Participation	50% carried interest At back-in Ecopetrol reimburses 50% of cost of *successful* wells
The government percentage of production increases with cumulative production above 60 MMBBLS.	

Cumulative Production, MMBBLS	Contractor Share of Production, %
0–60	50
60–90	45
90–120	40
120–150	35
Over 150	30

CONGO
Royalty/Tax (1993)

Area	
Duration	
Exploration	10 years
Production	30 years
Relinquishment	
Royalty	14.5%–19% Oil (Negotiated)
	9% Gas

The old royalty calculated on 80% of FOB price ranging from 6.5% at 1,000 BOPD to 15% for 100,000 BOPD. Gas 2%–5%.

Bonuses	250 MM CFA Francs at 30,000 BOPD
	625 MM CFA Francs at 75,000 BOPD
	may be capitalized and deducted
Depreciation	5-year straight-line decline
Taxation	55%
Domestic Market Obligation	Not to exceed 30% of contractor's oil
Ringfencing	Development area one ringfence
State Participation	50% Gvt reimburses contractor out of 75% of gvt net revenues with interest

EGYPT

Early 1980s
From: *Petroleum Company Operations and Agreements in Developing Countries,*
Raymond F. Mikesell, 1984

Area	Designated blocks
Duration	?
Relinquishment All except development leases after exploratory period	After each phase 25% of original area
Exploration Obligations For each commercial discovery, contractor must drill an additional other prospect.	Bid item
Royalty	0% EGPC pays royalties and taxes out of its share of profit oil
Signature Bonus	$1–$5 MM
Production Bonus	Yes
Cost Recovery	30% Limit
Depreciation	8-year SLD

Profit Oil Split (In favor of government)	Production, BOPD	Split, %
	Up to 90,000	80/20
	90,000–140,000	82.5/17.5
	140,000 +	85/15

Profit Gas Split	Negotiated
Taxation	Paid by government, therefore profit oil split is effectively an after-tax split
State Participation	50%/50% joint venture

EGYPT
1986 Standard Model

Area	Designated blocks

Duration
Exploration	8 years maximum, 3 phases
After discovery	1-year delineation period
Production	20 years

Relinquishment   After each phase 25% of original area
All except development leases after exploratory period

Exploration Obligations   Bid item
For each commercial discovery contractor must drill an additional other prospect.

Royalty	0%–10% Negotiated

Signature Bonus	$2–$4 MM

Production Bonus	$3 MM at 30,000 BOPD
	$5 MM at 50,000 BOPD
	$7 MM at 100,000 BOPD

Cost Recovery	40% Limit Offshore 30% Onshore

Depreciation	20%/year expl. & dev. costs

Profit Oil Split

(In favor of government)

Proposed in 1986 Production, BOPD	Split % Newer	Older
Up to 20,000	70/30	80/20
20,000–40,000	75/25	83/17
40,000 +	80/20	85/15

Most contracts continue to be based on standard flat 85%/15% split.

Profit Gas Split	Negotiated

Taxation	Paid by government, therefore profit oil split is effectively an after-tax split

State Participation	Nil

EQUATORIAL GUINEA
PSC ROR Contract

Area	125,000-acre grid blocks
Duration Exploration Production	5 years, 3 1-year/1-well extensions 30 years from commercial discovery 50 years for gas
Relinquishment	40% at end of 3rd year additional 20% before end of 5th year
Exploration Obligations	Drilling
Royalty	10%

Bonuses	Signature	$1 MM
	Discovery	$300,000 +
	First Oil Sales	$1 MM
	Production Bonuses	$2–$5 MM at 10,000–20,000 BOPD ±
Surface Rentals		$.50–$1.00 per hectare ±

Cost Recovery Limit	After paying royalty, no limit
Depreciation	25% per year (all costs)

Profit Oil Split (two examples shown here)
"Net Crude Oil" after royalties and cost recovery

Contractor's Pretax Real ROR, %	Contractor Share, %	Contractor's Pretax Real ROR, %	Contractor Share, %
up to 30	100	up to 20	100
30–40	60	20–40	80
40–50	40	40–60	50
over 50	20	over 60	10

Taxation	25% income tax
Ringfencing	
Domestic Market Obligation	If requested to do so, contractor sells to government a portion of net crude oil at market prices.
State Participation	
Other	

FRANCE
Late 1980s

Area	Minimum 175 km$_2$ after relinquishment
Duration Exploration Production	5 years + maximum of two 5-year extensions 5 years + two 5-year extensions for fields < 2.1 MMBBLS up to 50 years for large fields
Relinquishment	After: 5 years 50% Extension 12.5% (leaving 37.5%)
Exploration Obligations	
Royalty	Sliding Scale: Oil, % Up to 1,000 BOPD 0 1,001–2,000 6 2,001–6,000 9 6,001 + 12
Gas Royalty for production over 30 MMCFD = 5%	
Signature Bonus	
Cost Recovery	Concessionary System, 100%
Depreciation	
Taxation	50% income tax Tax deferral/depletion allowance 23.5% of gross or 50% of net income if spent on further exploration
State Participation	Nil

GABON
PSC (1989)

Area	0.1–1 million-acre blocks
Duration Exploration Production	 3 years + 2-year extension 20 years
Relinquishment	After: 3 years 25% 5 years 50%
Exploration Obligations	1–3 Well minimum
Royalty	Sliding Scale: Up to 10,000 BOPD 5% 10,001–20,000 10% 20,001–40,000 15% 40,001 + 20%
Signature Bonus	U.S. $0.5–$2 million
Production Bonus	Startup: U.S. $1.0 million 10,000 BOPD: 1.0 million 20,000 BOPD: 2.0 million
Cost Recovery	55% Limit 40% older contracts
Depreciation	5-year straight line
Profit Oil Split (In favor of government)	Production, BOPD Split, % Up to 5,000 65/35 5,000–10,000 70/30 10,001–20,000 73/27 20,001–30,000 75/25 30,001–40,000 80/20 40,000 + 85/15
Profit Gas Split	Negotiated
Taxation	56% income tax paid by government out of contractor share of profit oil—therefore profit oil split is effectively an after-tax split
Domestic Market Obligation	Up to 20% of profit oil sold at 75% of market price, otherwise pro rata
State Participation	10% Working Interest

GHANA
Royalty/Tax 1986 Vintage
ROR provision came later

Area

Duration
Exploration   7 years
Production   18 years (25 years including exploration)

Relinquishment   negotiated

Exploration Obligations

Royalty   12.5%

Bonuses   None or negotiated

Depreciation   20% per year (exploration & development)

Taxation   50% local income tax

Additional Profits Tax After-tax
[Post-1986]

ROR, %	Rate, %
Up to 15	50
15–25	60
over 25	70

Ringfencing

Domestic Market Obligation   At world prices

State Participation   GNPC carried for 10% through exploration
                           has option to increase participation
                           to majority interest during development

Other

INDIA PSC
Early 1990s

Area	?

Duration
Exploration
Production

Relinquishment

Exploration Obligations

Royalty	None
Bonuses	None
Cost Recovery	100% no limit
Depreciation	4 Years

Profit Oil Split

Investment Multiple Cumulative Net Cash Flow ÷ Exploration & Development Costs	Contractor Share, %
0–1.5	100
1.5–2.0	90
2.0–2.5	85
2.5–3.0	80
3.0–3.5	75
over 3.5	60

Taxation	50%
Ringfencing	Development costs are ringfenced by field. Exploration costs are not.
State Participation	30% Back-in

INDONESIA
Second Generation (Pre-1984)

Area	No restriction, designated blocks
Duration Exploration Production	 3 years 20 years
Relinquishment	25% or 100% of no discovery
Exploration Obligations	Multiwell commitments
Royalty	Nil
Signature Bonus	Various
Production Bonus	Many variations, each contract is different
Cost Recovery	No limit 20% Investment credit applies to facility, platform, pipeline costs; is recoverable but taxable
Depreciation	Oil — 7-year double declining balance going to straight line in year 5 Gas — 7-year declining balance switching to straight line in year 8
Profit Oil Split (In favor of government)	65.9091% / 34.0909%
Profit Gas Split (In favor of government)	20.4545% / 79.5455%
Taxation	56% income tax
Ringfencing	Each license ringfenced
Domestic Market Obligation	After 60 months production from a field, contractor receives 20¢/bbl for 25% of oil
State Participation	Up to 50%—Option seldom exercised

INDONESIA
Fourth Generation (Post 1988–89)

Area	No restriction, designated blocks
Duration Exploration Production	 3 years 20 years
Relinquishment	25% or 100% of no discovery
Exploration Obligations	Multiwell commitments
Royalty	Nil
Signature Bonus	Still exist, various
Production Bonus	Many variations each contract is different
Cost Recovery	80% limit because of 1st tranche petroleum of 20% 17% Investment credit applies to facility, platform, pipeline costs; is recoverable but taxable
Depreciation	Oil 25% declining balance with balance written off in year 5 Gas 10% declining balance with balance written off in year 8
Profit Oil Split (In favor of government)	 71.1574% / 28.8462%
Profit Gas Split (In favor of contractor)	 42.3077% / 57.6923%
Taxation	48% income tax
Ringfencing	Each license ringfenced
Domestic Market Obligation	After 60 months production from a field, contractor receives 10% of market price for 25% of oil
State Participation	Up to 50% in joint operating agreement contracts

230

IRELAND
Concession

Area	Designated blocks
Duration	
Relinquishment	
Exploration Obligations	
Royalty	None
Bonuses	
Cost Recovery Limit	100%, no limit
Depreciation	100%, all costs expensed
Taxation	25% tax on profits
Ringfencing	
Domestic Market Obligation	
State Participation	
Other	

IVORY COAST (Côte d'Ivoire
1988 Vintage PSC

Area	
Duration	
Exploration	5 years (3 periods: 2 + 1 + 2)
Production	25 years + additional period of 10 years
Relinquishment	50% after 2nd period (end of 3rd year)
Exploration Obligations	
Royalty	None with PSCs
Bonuses	Negotiated
	Model contract mentions $12 MM
Cost Recovery Limit	40%
Depreciation	
Profit Oil Split	

	Shallow Water	Deepwater
	< 1,000 m	> 1,000 m
	Contractor	Contractor
Production BOPD	Share, %	Share, %
Up to 30,000	52	60
30,001–50,000	48	56
50,001–100,000	38	54
100,001–120,000	32	54
more than 120,000	30	54

Taxation	50% maximum tax on profits
	Set at 34% in 1993
Ringfencing	
Domestic Market Obligation	Up to 15% or prorated share
	Contractor receives 15% of FOB price
State Participation	Varies between 10% and 60%. Fixed at 15%
	for deepwater (NOC Petroci)
	After 1993 reduced to 10% with state
	option for up to 30% after discovery.
Other	

KOREA
Concession

Area	
Duration Exploration Production	
Relinquishment	
Exploration Obligations	
Royalty	15%
Bonuses	
Cost Recovery Limit	100%, no limit
Depreciation	
Taxation	50%
Ringfencing	
Domestic Market Obligation	
State Participation	

MALAYSIA
Late 1980s, Early 1990s

Area	No restriction, designated blocks
Duration	
Exploration	3 years + 2-year extension
Development	2 years + 2-year extension
Production	15 years for oil/20 years for gas
Relinquishment	No interim relinquishment
Exploration Obligations	Seismic and multiwell commitments
Royalty	10%
	0.5% Research Cess
Signature Bonus	None (Older contracts had bonuses)
Production Bonus	None (Older contracts had bonuses)
Cost Recovery	50% limit for oil/60% for gas
Depreciation	10% year straight line

Profit Oil Split (In favor of government)	Production, BOPD	Split, %
	Up to 10,000	50/50
	10,001–20,000	60/40
	20,001 +	70/30
	All production in excess of 50 MMBBLS	70/30

Profit Gas Split (In favor of contractor)	For first 2 TCF	50/50
	After 2 TCF produced	70/30

Taxation	25% duty on profit oil exported
	(with 20% export tax exemption)
	45% petroleum income tax
Ringfencing	Each license ringfenced
Domestic Market Obligation	Nil
State Participation	Up to 15%

MALAYSIA
1994

Area	No restriction, designated blocks
Duration Exploration Development Production	3 years + 2-year extension 2 years + 2-year extension 15 years for oil/20 years for gas
Relinquishment	No interim relinquishment
Exploration Obligations	Seismic and multiwell commitments
Royalty	10% 0.5% Research Cess
Signature Bonus	None (Older contracts had bonuses)
Production Bonus	None (Older contracts had bonuses)
Cost Recovery	50% limit for oil/60% for gas
Depreciation	10% year straight line

Profit Oil Split
(In favor of government)

Production, BOPD	Split, %
Up to 10,000	50/50
10,001–20,000	60/40
20,001 +	70/30
All production in excess of 50 MMBBLS	70/30

Profit Gas Split
(In favor of contractor)

For first 2 TCF	50/50
After 2 TCF produced	70/30

Taxation	20% duty on profit oil exported (with 50% export tax exemption) 40% Petroleum income tax

Ringfencing   Each license ringfenced
Also, gas development costs recovered from gas production, and oil development costs recovered from oil production.

Domestic Market Obligation	Nil
State Participation	Up to 15%

235

MALTA
PSC 1988 Vintage

Area	
Duration Exploration Production	
Relinquishment	
Exploration Obligations	
Bonuses	Negotiable at 3 levels Production, BOPD 50,000 100,000 150,000
Royalty	None
Cost Recovery Limit	No Limit
Depreciation	All costs 25% per year
Profit Oil Split	Negotiable

	(Example)
Production, BOPD	Contractor's Share, %
Up to 50,000	75
50,000–100,000	70
over 100,000	60

Taxation	50% of contractors profit oil Bonuses are not cost recoverable, but they are tax deductible
Ringfencing	
Domestic Market Obligation	None
State Participation	None

MOROCCO
Concession (1983 License Round)

Area	at least 500 and less than 5,000 km²
Duration	
Exploration	4 years + 2–3 4-year renewals
Production	
Relinquishment	25% on first renewal
	25% on second renewal
	12.5% on third renewal
Exploration Obligations	

Royalty

Sliding Scale	Rate, %
Up to 1,000 BOPD	0
1,001–2,000	6
2,001–6,000	9
6,001–20,000	12
20,001 +	14

Total royalties may not exceed 12.5% of gross value of production

Bonuses	
Rentals	$2-6/1,000 acres/year initial period
	$3-12/1,000 acres/year after 1st renewal
	$12-25/1,000 acres/year after 2nd renewal
Depreciation	5-year SLD
Taxation	48% profits tax
	Special surtax after 4 years continued production in excess of 7,500 BOPD or 1 MMCFD gas. Tax is equal to the difference between total taxes assessed including royalties and rentals and 50% of net profits.
Ringfencing	?
Domestic Market Obligation	None
State Participation	None

MOROCCO
Royalty/Tax (1989 with 1986 incentives)

Area

Duration
   Exploration
   Production

Relinquishment

Exploration Obligations

Royalty	12.5% of gross value of production after first 4 MM tons (28 MMBBLS) have been produced
Bonuses negotiable not deductible for tax purposes	upon discovery also production bonuses
Depreciation	5-year DDB exploration capital 10-year SLD development capital
Taxation	52.8% effective tax rate

Special surtax based on profit/investment ratio

Surtax, %	Assumed Ratio
10	1.0
20	1.5
30	2.0
40	2.5
50	3.0

Ringfencing	?
Domestic Market Obligation	None
State Participation	35% carried through exploration

MYANMAR
First License Round 1989/1990

Area	No restriction, designated blocks
Duration Exploration Production	3 + 1 + 1 years 20 years
Relinquishment	25% + 25% or 100% if no discovery
Exploration Obligations (Initial Phase)	Negotiable U.S. $12–$88 million Averaged U.S. $20 million
Royalty	10% + 0.5% for research & training
Signature Bonus	U.S. $4.0–$7.5 million
Production Bonus	Discovery: U.S. $1.0 million 10,000 BOPD 2.0 million 30,000 BOPD 3.0 million 50,000 BOPD 4.0 million
Cost Recovery	40% limit
Depreciation	10%

Profit Oil Split
(In favor of government)

Production, BOPD	Split, %
Up to 50,000	70/30
50,001–100,000	80/20
100,001–150,000	85/15
150,001 +	90/10

Profit Gas Split
(In favor of government)

Production, MMCFD	Split, %
Up to 300	70/30
301–600	80/20
601–900	85/15
901 +	90/10

Taxation	30% income tax

Tax holiday first 3 years under Foreign Investment Law

Domestic Market Obligation	Pro rata: up to 20% of contractor's share of oil at U.S. $1/bbl
State Participation	Nil

SUMMARY OF PRODUCTION SHARING CONTRACT TERMS

MYANMAR		ONSHORE		OFFSHORE	
1992	New EP Blocks	IOR	RSF	<600ft	>600ft
Exploration Period	3 years	6-month study period 2 yrs to get increase	6-month study period 2 yrs to start production	3 years	3 years
Extensions	2 X 1			2 X 1 year	2 X 1 year
Development Period	20 years for each discovery	15 yrs from increase of production	20 yrs from start of production	20 years for each discovery	20 years for each discovery
Relinquishment	25% 1st extension period	0 or 100% after study period	0 or 100% after study period	25% 1st extension period	25% 1st extension period
Royalty	10%	10%	10%	10%	10%
Cost Recovery Limit	40%	40%	40% (MOGE recovers sunk costs 50/50)	50%	50%
Production Sharing Oil BOPD Gvt./Contractor%	30,000 65/35% 50,000 70/30 100,000 80/20 150,000 85/15 150,000+ 90/10	10,000 70/30% 20,000 75/25 30,000 80/20 30,000+ 85/15	10,000 70/30% 20,000 75/25 30,000 80/20 30,000+ 85/15	25,000 60/40% 50,000 65/35 100,000 80/20 150,000 85/15 150,000+ 90/10	25,000 60/40% 50,000 65/35 100,000 75/25 150,000 80/20 150,000+ 90/10
Production Sharing Gas MMCFD	180 65/35% 300 70/30 600 80/20 900 85/15 900+ 90/10			300 65/35% 600 75/25 900 85/15 900+ 90/10	300 60/40% 600 70/30 900 80/20 900+ 90/10
Signature Bonus	$5 MM	$5 MM after study period	$5 MM after study period	$5 MM +	$5 MM
Discovery Bonus	$1 MM	$500,000 data	$500,000 data	$1 MM	$1 MM
Production Bonus	10,000 $2 MM 30,000 3 MM 50,000 4 MM 100,000 5 MM 200,000 10 MM	$200,000 commerciality	$1 MM at commerciality	25,000 $2 MM 50,000 3 MM 100,000 4 MM 150,000 5 MM 200,00010 MM	25,000 $2 MM 50,000 3 MM 100,000 4 MM 150,000 5 MM 200,000 10 MM
Income Tax	30% 3-yr holiday	30% 3-yr holiday	30% 3-yr holiday	30% 3-yr holiday	30% 3-yr holiday
Training	$50,000 per yr	$50,000 per yr	$50,000 per yr	$50,000 per yr	$50,000 per yr
R&D	0.5% of profit	0.5% of profit	0.5% of profit	0.5% of profit	0.5% of profit
Domestic Oil Requirement	@ 60% of world price		20% of profit oil @ 10% price		
Other			10% finders bonus (royalty)		

Most terms are negotiable. This table represents general or proposed terms outlined by MOGE.

NEW ZEALAND, Concessionary
Proposed 1991 Crown Minerals Act (Awaiting ratification)

Area	No restriction, designated blocks
Duration Exploration Production	 5 years with 5-year extension Life of the field
Relinquishment	50% after 5 years
Exploration Obligations	Negotiable
Royalty (Hybrid)	5% ad valorem royalty (AVR) or 20% accounting profits royalty (APR) whichever is greater Previously the royalty was a flat 12.5% AVR
Signature Bonus	Negotiable No
Cost Recovery	No limit
Depreciation	20%
Taxation	33% income tax (resident companies) 15% withholding tax 38% income tax (nonresident companies)
State Participation	Nil (Previously was 11% carry through exploration phase.)

NIGERIA
PSC 1987, Ashland Contract

Area

Duration
  Exploration
  Production

Relinquishment

Exploration Obligations

Royalty   20%

Bonuses

Cost Recovery Limit   40%

Depreciation

Profit Oil Split (In favor of government)	Production, BOPD	Split, %
	Up to 50,000	65/35
	over 50,000	70/30

1986 terms guaranteed $2/bbl profit margin on equity crude.

Taxation	85% 65% during cost recovery "while amortizing preproduction costs"

Ringfencing

Domestic Market Obligation

State Participation

Other   Some uplifts/investment credits

NIGERIA
PSC New 1994 Terms

Area

Duration
  Exploration
  Production

Relinquishment

Exploration Obligations

$24 million first 3 years
$30 million next 3 years
$60 million ten additional years
Former requirement: $176 MM over 10 years

Royalty	Water Depth, m	Rate, %
	up to 200	16.667
	200–500	12
	500–800	8
	800–1000	4
	> 1000	0

Bonuses

$2 million @ 10,000 BOPD
$2 million @ 50,000 BOPD

Cost Recovery Limit

? Under old contracts the limit was 40%

Depreciation

Profit Oil Split (In favor of government)	Production, BOPD	Split, %
	Up to 100,000	55/45
	100,001–200,000	60/40
	over 200,000	62/38

Taxation   50%
Down from 85% under older contracts, which had lower 65% rate during cost recovery period.

Ringfencing

Domestic Market Obligation

State Participation

Other   50% investment credit

NORWAY
Concession

Area	
Duration Exploration Production	30 Years Field Specific
Relinquishment	
Exploration Obligations	
Royalty	0 (Post-1986) Prior to 1986 royalty ranged from 8%–14%
Bonuses	None
Cost Recovery Limit	100%
Depreciation	6-years SLD Beginning in a year of investment. Prior to 1986 "when placed in service"
Taxation	28% income tax 30% special tax

The basis of the special tax is *free income,* which is similar to ordinary income tax basis but includes additional deduction for 5% uplift on development capital costs.

Ringfencing	Not in upstream end
Domestic Market Obligation	None
State Participation	Statoil has option on up to 80% working interest—no carry. Prior to 1986 government was carried through exploration.
Other	Prior to 1986, 5% uplift on dev. cap. ex. for 6 years, abolished. 0.7% Tax capital tax = ad valorem tax on book value of investments 15% production credit (deduction)

PAKISTAN
Concession (Mid-1980s vintage)

Area	Maximum 125 km$_2$
Duration	20 yrs onshore, 25 offshore
Relinquishment	25% after 4 years + 25% after 2 more
Exploration Obligations	
Royalty	12.5% less annual rentals
Bonuses	OXY had $1 MM at commercial production $1.5 MM at 5,000 BOPD $3 MM at 25,000 BOPD $5 MM at 50,000 BOPD or BOE Equivalent (6:1)
Depreciation	
Taxation	50% 55% Maximum
Ringfencing	
Domestic Market Obligation	Pro rata
State Participation	Government had option to acquire 25% Working interest in OXY block. 40% Working interest in Badin block.

PAPUA NEW GUINEA
Concession (ROR)

Area	Graticular blocks 5 minutes longitude X 5 minutes latitude 9 km X 9 km graticules Licenses may be 60–200 blocks (approximately 1.25–4 million acres)
Duration Exploration Production	 Under petroleum prospecting license (PPL) 6 years + 5-year extension for 50% of area if work program complete Under petroleum development license (PDL) 25 years with 20-year extension
Relinquishment	Surrender 25% after first 2 years
Exploration Obligations	Negotiated
Royalty	1.25%
Bonuses	Negotiated
Cost Recovery Limit	No limit

Company allowed to recover its investment plus agreed rate of interest U.S. AAA Bond rate + 5%

Depreciation	8-year SLD. Some accelerated (4-year SLD) allowed if target income is not met.
Taxation	50% Basic Petroleum (income) Tax (BPT)

Exempt if target income test is not met (25% of investment)

50% Additional Profits Tax (APT)

Resource Rent Tax based on 27% ROR threshold test.

Ringfencing

Domestic Market Obligation

State Participation	22.5% carried through exploration State share of Development costs paid out of State share of production.

Other

PHILIPPINES
Risk Service Contract early 1990s

Area	Designated blocks

Duration	
Seismic Option	1 year
Exploration	10 year maximum
Production	30 years

Relinquishment	

| **Exploration Obligations** | Negotiable |
| | Two-well option after seismic |

Royalty*	−7.5% (goes to contractor group)
	Depends upon level of Filipino
*Filipino Participation	ownership up to 30% onshore
Incentive Allowance (FPIA)	up to 15% in deepwater qualifies
	for full 7.5% (FPIA)

Filipino Participation, %	FPIA, %
Up to 15%	0
15–17.5	1.5
17.5–20	2.5
20–22.5	3.5
22.5–25	4.5
25–27.5	5.5
27.5–30	6.5
30 or more	7.5

Signature Bonus	Negotiable

Production Bonus	No

Cost Recovery	70% limit

Depreciation	10%

| **Profit Oil Split** | 60%/40% |
| (In favor of Government) | Contractor's 40% is a "service fee" |

Taxation	No, paid out of gvt share

| **Ringfencing** | Cost recovery allowed on two or more |
| | deepwater blocks |

Domestic Market Obligation	Prorata

State Participation	Nil (FPIA)

SPAIN
Royalty/Tax

Area	10,000 to 40,000 hectares 24,700 to 98,800 acres
Duration Exploration Production	 30 with two 10-year extensions
Relinquishment	
Exploration Obligations	
Royalty	None
Bonuses	Negotiable, unlikely
Depreciation	Exploration and intangible costs amortized over 4 years SLD Most tangible costs capitalized 4 years SLD Platforms 8 years SLD Pipelines 5 years SLD
Taxation	40% income tax
Ringfencing	
Domestic Market Obligation	
State Participation	None
Other	10% Investment credit on tangible capital costs 25% Depletion allowance on gross revenues if it is reinvested in Spain, but limited to 40% of taxable income

SYRIA
PSC

Area	

Duration	
Exploration	
Production	

Relinquishment	

Exploration Obligations	

Royalty	11% 1985 Pecten Group royalty is 12.5%

Bonuses	Production Bonuses
	50,000 BOPD
	100,000 BOPD
Not recoverable	200,000 BOPD

Cost Recovery Ceiling	35% of gross production less royalties

1985 Pecten Group Cost Recovery Ceiling 25% and unused cost oil goes directly to government

Depreciation	Exploration capital and operating costs expensed. Development costs: 5 years SLD

Profit Oil Split

————— Sample Ranges —————

Production, BOPD	Example 1, %	Example 2, %	Pecten 1985, %
Up to 25,000	22.5	25	21
25,001–50,000	21.36	24	21
50,001–100,000		20	19
100,001–200,000		18	19
over 200,000	13.35	15	15

Taxation	Taxes paid by government on behalf of contractor

Ringfencing	

Domestic Market Obligation	

State Participation	None

Other	

THAILAND
Royalty/Tax Contract early 1990s

Area	Designated blocks
Duration	
Relinquishment	
Exploration Obligations	

Royalty		Rate, %
	up to 2,000 BOPD	5
	2,000–5,000	6.25
	5,000–10,000	10
	10,000–20,000	12.5
	over 20,000	15

Signature Bonus	Yes, $2–$5 million
Production Bonus	No
Depreciation for preproduction/postproduction expenses	5 Years for tangibles, intangibles 10 Years
Taxation	50% Income tax

Supplemental Tax Special remuneratory benefit (SRB)
Progressive rate from 0%–75%
Based generally upon ratio of annual petroleum profit ÷ by cumulative depth in meters
of all wells drilled in the block plus a constant—100,000 m, for example.

$/M	SRB Rate
Less than $200	0%
200–580	1% per $10/m
580–1,340	40% + 1% per $40/m
1,340–3,650	60% + 1% per $154/m
over 3,650	75%

Assuming 24 Thai Bhat per U.S. $1.00.
The more drilling, the lower the tax: this is an unusual one.

Ringfencing	?
Domestic Market Obligation	
State Participation	?

TIMOR GAP
Zone of Cooperation, 1991–92 License Round
PSC Jointly Administered by Indonesia & Australia

Area	Main blocks in Zone of Cooperation A (ZOCA) comprise 20–40 subblocks at 10 km$_2$ each
Duration Exploration	6 years with option for 4-year extension With development contract automatically extends to 30 years
Committed Expenditures Exploration	First year seismic only $1–$4 MM Second year 0–2 wells $.5–$8 MM Third year 1–3 wells $.5–$21 MM 4th–6th years 1–4 wells $6–$30 MM
Relinquishment	25% after 3 years; another 25% after 6th year
Royalty	None
Bonuses	
Cost Recovery Limit	90% Effective limit for 1st 5 years* 80% Effective limit thereafter*

*10% first-tranche petroleum (FTP) (similar to Indonesian FTP) after 5 years production reverts to 20% FTP

Depreciation	5-year SLD

Profit Oil Split

Production, BOPD	Contractor Share, %
Up to 50,000	50
50,001–150,000	40
150,001–200,000	30
Natural Gas	50

Taxation	48% Effective tax rate (similar to Indonesia) Comprised of 35% income tax and 20% withholding tax

Companies will lodge income tax returns with both countries. In each country a 50% tax rebate will be given.

Ringfencing	
Domestic Market Obligation	Similar to Indonesian DMO

25% of pretax profit oil (After 60 months, 10% of market price)

State Participation	None
Other	17% IC on eligible costs similar to Indonesian IC 127% investment credit (IC) for deepwater

TUNISIA
New Hydrocarbon Laws, 18 June 1990
Concession

Area

Duration
Exploration
Production

Relinquishment

Exploration Obligations

Royalty	R Factor	Oil, %	Gas, %
	< .5	2	2
	.5–.8	5	4
	.8–1.1	7	6
	1.1–1.5	10	8
	1.5–2.0	12	9
	2.0–2.5	14	10
	2.5–3.0	15	11
	3.0–3.5	15	13
	3.5 +	15	15

R factor = accrued net earnings/accrued total expenditures

Bonuses

Depreciation   30% per year (all investments)

Taxation

R Factor	Income Tax Rate, %
< 1.5	50
1.5–2.0	55
2.0–2.5	60
2.5–3.0	65
3.0–3.5	70
3.5 +	75

Ringfencing

Domestic Market Obligation   Pro rata up to maximum 20%
DMO price = FOB price −10%

State Participation
Contractor recoups state share
of exploration costs out of
20% of state share of revenues.

R Factor	Sample Level, %
< 1.5	45
1.5 +	50

TURKMENISTAN
"Joint Enterprise" Contracts
Summarized from *Oil & Gas Journal,* Vol. 91, No. 6, Feb. 8, 1993 (pp. 38–39)

Area	

Duration	25 years with optional 10-year extensions

Relinquishment	

Obligations	Block
	II $60 MM 5 years
	III $50 MM 5 years
	IV $50 MM 5 years

Royalties (Sliding scale)

Blocks II & III, BOE/J	Rate	Block IV, BOE/J	Rate
Up to 3,649	0%	Up to–7,299	0%
3,656–7,299	2	7,300–21,899	2
7,300–10,949	5	21,900–36,499	5
10,950–18,249	7	36,500–51,099	7
18,250 +	15	51,100 +	15

Bonuses		Amount		
Block	Minimum	Bid	Reserves	Group
II	$15 MM	$15.25 MM	230 MMBBLS + 1.87 TCF	Larmag/Noble
III	$20 MM	$20 MM	642 MMBBLS + 2.16 TCF	Eastpac/TMN
IV	$30 MM	$30 MM	230 MMBBLS + .89 TCF	Bridas

Depreciation	?

Production Sharing	
	Block
	II 50% 50% split
	III 10% 90% in favor of the government
	IV 30% 70% in favor of the government
	These quoted percentages are possibly "after-tax"?

Taxation	35% on joint venture profits Guaranteed against increases

Ringfencing	?

Domestic Market Obligation	?

State Participation	50% JV fully carried?

Other	Net operating losses (NOL) carried forward 5 years

UNITED KINGDOM
Concession, Early 1990s

Area	Designated blocks
Duration Exploration Production	18 years Field specific
Relinquishment	
Exploration Obligations	Bid, negotiated
Royalty	Nil
Bonuses	None
Cost Recovery Limit	100% no limit
Depreciation	25% declining balance
Taxation	33% income tax 75% petroleum revenue tax (PRT) on net revenues after capital costs are recovered. Some limits for marginal fields. Also free oil allowance against PRT
Ringfencing	Each license is ringfenced For PRT each *field* is ringfenced
Domestic Market Obligation	None
State Participation	None
Other	35% uplift on some capital costs

UNITED KINGDOM, Fiscal
Summary of 1983–84 changes:

	Before 1985	After 1985
INCOME	$100	$100
ROYALTY Abolished for projects approved after April 1992	12.5%	0%
Net Revenue	87.5	100
PRT (75%) Phaseout by end of 1986	65.62	0
	21.88	100
CORPORATE TAX	11.38 (52%)	35 (35%)
Contractor Share	10.5	65
GOVERNMENT TAKE	89.5%	35%

Current corporate tax rate in the UK is 33% which yields a "pure" 67/33% split in favor of contractor group for fields developed since1982.

Amount of oil exempted from PRT doubled to 1 million metric tons per year ≈ 20,000 BOPD. Cumulative limit ≈ 10 million metric tons (73 MMBBLS).

In effect, on a field of 20,000 BOPD or less, only tax is corporate tax of 35%. No APRT, PRT, or royalty.

HISTORY

Petroleum Production Act of 1934 Fiscal framework vested oil and gas ownership with the Crown.

Oil Taxation Act of 1975 Introduction of PRT (initially 45%) based upon profits of each individual field. "Ringfencing" introduced as Part II of the Act prevents oil companies from offsetting profits with other losses within the company. However, N. Sea losses can be set against profits from other activities in a company.

(PRT gradually increased)
1975   Government increased take to 76.9%
1979   Government increased take to 83.2%
1980   Government increased take to 87.4%

1981 *Supplementary Petroleum Duty* 20% tax, which was withdrawn after two years. This had increased government take to 90.3%. PRT rate in 1981 was increased to 75%.

APRT ACT 1986 Between 1983 and 1986 advance payments of PRT (APRT) were required in early stage of field, even though no PRT may have been required at that time. The last payment of APRT was for the chargeable period ending December 1986. Up to then, the PRT liability had not been sufficient to absorb all the APRT paid. The Act provided for immediate repayment of up to £15M (15 million pounds sterling per field participant).

FINANCE ACT of 1983 Exploration and appraisal drilling expenses were available for immediate PRT relief; oil allowance for new fields doubled; new oil fields exempted from royalty.

FINANCE ACT of 1987 Up to 10% of development costs on certain new fields can be set against PRT liabilities in existing fields.

PRESENT SYSTEM
ROYALTY: 12.5%   for licenses issued in 1st–4th rounds transportation costs may be deducted.

No transportation costs may be deducted for 5th and subsequent licensing rounds.

No royalty for onshore or offshore fields which received development approval after 1 April 1982.

Discretionary Royalty Relief: based upon the Petroleum and Submarine Pipelines Act of 1975, the Secretary of State for Energy with consent of the Treasury can refund royalties in whole or in part to provide incentive to develop or to continue to produce.

CORPORATE TAX (CT): 35% on net revenues after deduction of royalty, PRT, and ringfenced expenses.

PETROLEUM REVENUE TAX (PRT): 75% of profits less various reliefs. Also charged on tariff receipts

1. Gas pre-30 June 1975 exempt.
2. Up to 10% of costs to develop certain new fields can be set against PRT.
3. Provisions for losses on abandoned fields.
4. Expenses allowable against PRT:

   a. License Royalties
   b. Capital Expenditure + 35% Uplift on certain Expl & Dev costs prior to payback. Intended to compensate for interest and costs of financing—not deductible for PRT purposes.
   c. Oil Allowance exempts from PRT fixed amount of oil or gas subject to a cumulative total.

Section 8 OTA 1975   250,000 metric tons (5,000 BOPD)
          Cumulative limit 5 MM tons (36 MMBBLS)
          all fields, pre-1 April 1982

Section 36 FA 1983   500,000 metric tons (10,000 BOPD)
          Cumulative limit 5 MM tons (36 MMBBLS)
          offshore outside S. Basin after 4/82
Section 138 FA 1988 125,000 metric tons (2,500 BOPD)
          Cumulative limit 2.5 MM tons (18 MMBBLS)
          Onshore & S. Basin fields after 4/82

Safeguard Provision: if during periods before payback and half the periods after, the PRT charge would return on a field before corporation tax (CT) to less than 15% of cumulative "upliftable" expenditure (measured on historical cost), the charge (PRT) is to be cancelled.

UZBEKISTAN
Joint Ventures
1st License Round 1993

Area	10 designated blocks

Duration	
Exploration	7 years
Production	23 years

Relinquishment	25% after 4 years and 25% each year thereafter

Minimum Work Commitments	3 wells on 9 blocks (2 on one block) 1,000–2,000 km seismic, Average 1,600

Royalty	Bid item, maximum 10% (Fixed or sliding scale)

Minimum Signature Bonus Production Bonuses	Average $1.1 million ($0–$2 MM) (bid item at 25, 50, and 100 MBOPD) Geodata Packages $30–$60,000

Cost Recovery Limit	Bid Item Maximum 60% Before_____and after_____recovery of initial costs

Depreciation	5-year SLD

Profit Oil/Gas Splits

Oil BOPD	Bid Items Split %	Gas MMM₃/Day	Split %
Up to 10,000	___/___	up to 5	___/___
10,001–25,000	___/___	5–10	___/___
25,001–50,000	___/___	10–15	___/___
50,001–75,000	___/___	over 15	___/___
75,001 +	___/___		

Proposed level of contractor share to range from 20% to 30%

Taxation	Income Tax 18% with 30% foreign ownership as low as 10% with > 30% foreign ownership as high as 35% with < 30% foreign ownership Possible 5-year holiday starting with operations Export Tax 10% (Repatriation of profits tax) VAT 25% on goods & services except G&G Property Tax 1%, 2-year exemption

The following yet to be determined by the Cabinet of Ministers
Tax on raw materials, excise charges, land tax, and
payment for use of natural resources tax

Ringfencing	

Domestic Market Obligation	

VIETNAM
From *PetroMin Magazine*, July 1991

Area	No restriction, designated blocks
Duration Exploration Production	 3 + 1 + 1 years 20 years
Relinquishment	25% to 35% or 100% if no discovery
Exploration Obligations (Initial Phase)	Minimum U.S. $50–$60 million or 3 exploration wells
Royalty	Nil
Signature Bonus	U.S. $ 0.5 million
Production Bonus	Discovery: U.S. $ 2.5 million 50,000 BOPD: 2.5 million 100,000 BOPD: 3.5 million 150,000 BOPD: 4.0 million
Cost Recovery	40% limit or 16% plus entitlement to purchase 29% to 40% of oil at discounted prices
Depreciation	Not clear

Profit Oil Split
(In favor of government)

Production, BOPD	Split, %
Up to 15,000	67/33
15,001–30,000	72/28
30,001–70,000	76/24
70,001–100,000	80/20
100,001 +	Negotiable

Profit Gas Split	Negotiable
Taxation	Taxes paid by Petrovietnam profit oil split is effectively an "after-tax" split
Ringfencing	Each license ringfenced
Domestic Market Obligation	Nil
State Participation	Nil

VIETNAM
Fina/Shell Contract, 16 June 1988

Area	Blocks 112, 114, 116

Duration
Exploration   5 years + 6-month extension for drilling
Production   25 years + 5-year extension

Relinquishment   25% at end of 3rd year
25% at end of 4th year
The whole of remaining areas relinquished after exploration period except development areas.

Exploration   First 3 years
Obligations
     10,000 km seismic or at least $6.5 MM
     3 wells or at least $8 MM per well
   Fourth year   5,000 km seismic or at least $4 MM
     2 wells or at least $8 MM per well
   Fifth year   2 wells or at least $10 MM per well
The last wells in 4th and 5th years conditional upon results of first well. At least one of first 4 wells to be at least 3,500 m deep.

Royalty   Nil

Signature Bonus   U.S. $ 1 million (deductible)

Production Bonus   Startup: U.S. $ 1.0 million
(not deductible)   50,000 BOPD: 2.0 million
     75,000 BOPD: 3.0 million

Cost Recovery   38.5% limit (60% for Gas)
Depreciation   Not clear

Profit Oil Split
Gas Same (6:1)

Production, BOPD	Before Payout, %	After Payout, %	After Threshold, %
Up to 50,000	40	36	32
50,001–60,000	37.5	33.75	30
60,001–70,000	35	31.5	28
70,001–80,000	32.5	29.25	26
80,001–90,000	30	27	24
90,001–100,000	25	22.5	20
100,001 +	20	18	16

Threshold volume is based upon an a cumulative production level.

Taxation   Taxes paid by Petrovietnam
Profit oil split is effectively an "after-tax" split

Ringfencing   Each license ringfenced

Domestic Market Obligation   Government has option to take all
at market price

State Participation   15% ?

N. YEMEN
Hunt Onshore PSC 1981

Area	12,600 km₂ (3 million acres ±)
Duration Exploration Production	 20 years
Relinquishment	
Exploration Obligations	
Royalty	None
Bonuses	
Cost Recovery Limit	30%
Depreciation	
Profit Oil Split	85%/15% in favor of the government
Taxation	
Ringfencing	
Domestic Market Obligation	
State Participation	
Other	

Some of the newer contracts have roughly 40% cost recovery limit and 70%/30% profit oil split in favor of the government.

Appendix B

Perspectives on Economic Rent

Rent theory is the foundation of petroleum taxation. The following discussions may be helpful.

David Ricardo, *The Principles of Political Economy and Taxation*, 1911 (1976 edition), Chapter II On Rent, page 33:

> Rent is that portion of the produce of the earth which is paid to the landlord for the use of the original and indestructible powers of the soil. It is often, however, confounded with the interest and profit of capital, and, in popular language, the term is applied to whatever is annually paid by a farmer to his landlord. . . .
>
> Adam Smith sometimes speaks of rent in the strict sense to which I am desirous of confining it, but more often in the popular sense in which the term is usually employed. He tells us that the demand for timber, and its consequent high price, in the more southern countries of Europe caused a rent to be paid for forests in Norway which could before afford no rent. Is it not, however, evident that the person who paid what he thus calls rent, paid it in consideration of the valuable commodity which was then standing on the land, and that he actually repaid himself with a profit by the sale of the timber?

If indeed, after the timber was removed, any compensation were paid to the landlord for the use of the land, for the purpose of growing timber or any other produce, with a view to future demand, such compensation might justly be called rent, because it would be paid for the productive powers of the land; but in the case stated by Adam Smith, the compensation was paid for the liberty of removing and selling the timber, and not for the liberty of growing it. He speaks also of the rent of coal mines, and of stone quarries, to which the same observation applies—that the compensation given for the mine or quarry is paid for the value of the coal or stone which can be removed from them, and has no connection with the original and indestructible powers of the land. This is a distinction of great importance in an inquiry concerning rent and profits; for it is found that the laws which regulate the progress of rent are widely different from those which regulate the progress of profits, and seldom operating in the same direction. . . .

In the future pages of this work, then, whenever I speak of the rent of land, I wish to be understood as speaking of that compensation which is paid to the owner of land for the use of its original and indestructible powers.

This specific concern over profit sharing acquires a distinctive importance from the element of economic rent in the exploitation of natural resources. Put simply, this rent element comprises that part of the marginal gain derived from working a given mineral deposit over its next cheapest alternative.

Ministry of Commerce (1992) Crown Minerals Act 1991: Evaluation of Allocation and Pricing Regimes, Resources Policy, Energy and Resources Division, Ministry of Commerce, Wellington, New Zealand:

> . . . the Ricardo model of rent generation relates to plots of similar areas and implicitly assumes the application of a common development and production technology. This means that the concept of economic rent has to be widened to take account of variations in the scale of production and the form of technology that is utilized. In this wider context economic rent can be defined, as the "residual value which exists after all factors of production, including capital have returned their opportunity costs.

Raymond F. Mikesell, *Petroleum Company Operations and Agreements in the Developing Countries*, Washington, D.C., 1984.

Chapter 4, page 30:

> . . . while the third objective (maximizing host government's revenue from oil lands) has to do with the division of the economic rent between the host government and the petroleum company or companies. By economic rent we mean the surplus of revenue over full economic costs of producing oil.

Page 46:

> Accounting profits vary widely over time and do not constitute a proper measure of economic

rent since they do not reflect full economic cost. The only proper measure of pure economic rent is the surplus over the net return to the investor that yields an IRR equal to the opportunity cost of capital. Moreover, if an investment is to be made, this IRR must be adjusted for risk.

Kamal Hossain, *Law and Policy in Petroleum Development,* 1979, page 84:

. . . economic rent (is) the difference between total revenues and total cost, including the costs of management and an appropriate risk premium.

Zuhayr Mikdashi, *The International Politics of Natural Resources,* Ithaca and London, Cornell University Press, 1976.

. . . economic rent accruing to a given mineral deposit is equal to the difference between the cost (including normal return on the required capital, but not including "user cost") of producing that deposit and the cost of producing a marginal deposit.

Anthony Burris and William Robson, "Evaluation of Royalty Bid and Profit Share Bid Bidding Systems." Published under the auspices of the U.S. Department of Energy, *Exploration and Economics of Petroleum Industry,* Volume 20, Institute of Petroleum, 1981:

According to a predetermined schedule, the Department of the Interior (of the U.S.) offers for sale oil and gas leases on OCS tracts. These leases transfer the rights to exploit the oil and gas resources. These rights then have an economic value and reflect the bidder's evaluation of the economic rent he might gain by winning the lease. The economic rent to the bidder is the difference between the value of production and the resource extraction costs and is the amount the Government as lessor should seek to capture as societal value of the resource. This recognizes that all long-run costs such as returns on investment and risk premiums have been netted out.

Paul Davidson, Laurence Falk, and Loesung Lee, "The Relations of Economic Rents and Price Incentives to Oil and Gas Supplies." *Studies in Energy Tax Policy,* Ballinger Publishing Company, 1975. (Brannon, G. ed.), page 119:

To the extent that each oil and gas property has no alternative use, payments to the landowner in the form of leasehold bonuses, royalties, and for many foreign properties, taxes are essentially economic rents.

Khong Cho Oon, *The politics of oil in Indonesia: Foreign company-host government relations,* 1986, page 17:

This specific concern over profit sharing acquires a distinctive importance from the element of economic rent in the exploitation of natural resources. Put simply, this rent element comprises that part of the marginal gain derived from working a given mineral deposit over its next cheapest alternative.

Host governments naturally wish to secure as much as possible of the value of the economic rent, and this desire has led most of them to lay down the concept of a national claim to ownership of their minerals.

APPENDIX C

ABBREVIATIONS AND ACRONYMS

AAA	American Arbitration Association
$/BBL	Dollars per Barrel
$/BOE	Dollars per Barrel of Oil Equivalent
$/BOPD	Dollars per Barrel of Oil per Day
$/MCF	Dollars per Thousand Cubic Feet of gas
$/MCFD	Dollars per Thousand Cubic Feet of gas per day
ADB	Asian Development Bank
ADV	Ad Valorem Tax
AFE	Authorization for Expenditure
AIG	American International Group
AIGPRI	American International Group Political Risk, Inc.
API	American Petroleum Institute
APR	Accounting Profits Royalty
ASEAN	Association of Southeast Asian Nations
AVR	Ad Valorem Royalty
B	Billion
B/CD	Barrels per Calendar Day (refinery: 365 days)
B/SD	Barrels per Stream Day (usually 330 days)
BBL	Barrel (crude or condensate), 42 U.S. Gallons
BCF	Billion Cubic Feet of gas
BCPD	Barrels of Condensate Per Day
BIT	Bilateral Investment Treaty
BOE	Barrels of Oil Equivalent (see COE)
BOEPD	Barrels of Oil Equivalent Per Day
BOPD	Barrels of Oil Per Day
BtU	British Thermal Unit
BWPD	Barrels of Water Per Day
CAPEX	Capital Expenditures
CAPM	Capital Asset Pricing Model

CEO	Chief Executive Officer
CFC	Controlled Foreign Corporation
CIF	Cost, Insurance, and Freight
CNG	Compressed Natural Gas
COE	Crude Oil Equivalent (also called BOE)
COO	Chief Operating Officer
CPE	Centrally Planned Economy
CR	Cost Recovery
DCF	Discounted Cash Flow
DCFM	Discounted Cash Flow Method (not common)
DD&A	Depreciation, Depletion, and Amortization
DDB	Double Declining Balance Method
DMO	Domestic Market Obligation
DO	Domestic Obligation
Dwt	Deadweight (as in Dwt Tonnage = long tons)
E&P	Exploration & Production
EBO	Equivalent Barrels of Oil (See BOE and COE)
EIA	Energy Information Administration
EMV	Expected Monetary Value
EOR	Enhanced Oil Recovery (see IOR)
EPIC	Engineering, Procurement, Construction, and Installation
EV	Expected Value (see EMV)
FASB	Financial Accounting Standards Board
FC	Full Cost Accounting
FCPA	Foreign Corrupt Practices Act
FIFO	First-In First-Out
FMV	Fair Market Value
FOB	Free On Board
FOGEI	Foreign Oil and Gas Extraction Income
FORI	Foreign Oil Related Income (downstream)
FPIA	Filipino Participation Incentive Allowance
FSU	Former Soviet Union
G&A	General and Administrative Expenses

G&G	Geological and Geophysical
GAAP	Generally Accepted Accounting Principles
GAO	General Accounting Office
GM	General Manager
IC	Investment Credit
ICC	International Chamber of Commerce
ICJ	International Court of Justice
ICSID	International Centre for the Settlement of Investment Disputes
IDC	Intangible Drilling and Development Costs
IEA	International Energy Association (Associated w/ OECD)
IMF	International Monetary Fund (UN)
INA	Insurance Company of North America
IOR	Improved Oil Recovery (same as EOR)
IRR	Internal Rate of Return
ITC	Investment Tax Credit
JOA	Joint Operating Agreement
JOB	Joint Operating Body
L/C	Letter of Credit
LDC	Less Developed Country
LIBOR	London Interbank Offered Rate
LIFO	Last-In First-Out
Lloyds	Lloyds of London
LNG	Liquified Natural Gas
LP	Limited Partnership
LPG	Liquid Petroleum Gas
Ltd	Limited Liability (British Corporation)
M	Thousand
MM	Million
MBBLS	Thousand Barrels
MBO	Management Buy Out
MCFD	Thousand Cubic Feet of gas per Day
MIGA	Multilateral Investment Guarantee Agency

MMBBLS	Million Barrels
MMCF	Million Cubic Feet of gas
MNC	Multinational Corporation
NOC	National Oil Company
NOL	Net Operating Losses
NGL	Natural Gas Liquids
NPV	Net Present Value
NRI	Net Revenue Interest
OAPEC	Organization of Arab Petroleum Exporting Countries
OECD	Organization for Economic Cooperation and Development
OIDC	Oil Importing and Developing Countries
OPEC	Organization of Petroleum Exporting Countries
OPEX	Operating Expenses
OPIC	Overseas Private Investment Corporation
ORI	Overriding Royalty Interest (Same as ORRI)
ORRI	Overriding Royalty Interest (Same as ORI)
PDP	Proved Developed Producing (Oil & Gas Reserves)
PE	Price Earnings Ratio
PLC	(British) Public Limited Company
PO	Purchase Order
PSA	Production Sharing Agreement (same as PSC)
PSC	Production Sharing Contract (same as PSA)
PUD	Proved Undeveloped (Oil & Gas Reserves)
PV	Present Value
PVP	Present Value Profits
RI	Royalty Interests
RLI	Reserve Life Index
ROA	Return on Assets
ROC	Return on Capital
ROE	Return on Equity
ROI	Return on Investment
ROR	Rate of Return

RRA	Reserve Recognition Accounting
RRR	Resource Rent Royalty
RRT	Resource Rent Tax
RSA	Risk Service Agreement (not common—same as RSC)
RSC	Risk Service Contract (not common—same as RSA)
SE	Successful Efforts Accounting
SEC	Securities and Exchange Commission
Semi	Semisubmersible Drilling/Production Vessel
SFAS	Statement of Financial Accounting Standards
STB	Stock Tank Barrel
STOIP	Stock Tank Oil In Place
TAC	Technical Assistance Contract
TCF	Trillion Cubic Feet of Gas
TCM	Technical Committee Meeting
TEA	Technical Evaluation Agreement
TLCB	Tax Loss Carry Back
TLCF	Tax Loss Carry Forward
TNC	Trans-National Corporation
UNCITRAL	United Nations Committee on International Trade Law
VAT	Value-added Tax
WI	Working Interest
WPT	Windfall Profits Tax

APPENDIX D

WORLDWIDE PRODUCTION STATISTICS (1992)

	Country	Producing Oil Wells	Average Daily Production, BOPD	Average per Well, BOPD
1	Abu Dhabi	993	1,842,000	1,855
2	Algeria	1,446	1,161,159	803
3	Angola	486	545,734	1,123
4	Argentina	8,402	554,000	66
5	Australia	1,081	533,000	493
6	Austria	1,149	23,091	20
7	Bahrain	352	36,747	104
8	Bangladesh	25	1,064	42
9	Barbados	102	1,308	13
10	Benin	8	3,000	375
11	Bolivia	332	21,180	64
12	Brazil	6,249	625,670	100
13	Brunei	738	171,866	233
14	Cameroon	193	113,700	589
15	Canada	40,667	1,245,294	31
16	Chile	355	15,778	44
17	China	49,700	2,835,000	57
18	Taiwan	84	1,160	14
19	Colombia	2,905	438,000	151
20	CIS	148,990	8,949,000	60
21	Congo	396	152,841	386
22	Croatia	919	38,860	42
23	Denmark	120	156,889	1,307
24	Dubai	151	402,000	2,662
25	Ecuador	999	321,000	321

	Country	Producing Oil Wells	Average Daily Production BOPD	Average per Well, BOPD
26	Egypt	1,015	873,000	860
27	France	609	58,031	95
28	Gabon	329	300,000	912
29	Germany	2,308	65,000	28
30	Ghana	—	1,800	1,800
31	Greece	13	13,721	1,055
32	Guatemala	16	5,616	351
33	Hungary	1,776	37,333	21
34	India	2,681	546,970	204
35	Indonesia	8,047	1,504,558	187
36	Iran	688	3,455,000	5,022
37	Iraq	820	425,000	518
38	Israel	11	184	17
39	Italy	234	83,000	355
40	Ivory Coast	12	1,000	83
41	Japan	274	10,600	39
42	Jordan	4	58	15
43	Kuwait	295	880,000	2,983
44	Libya	1,092	1,492,000	1,366
45	Malaysia	561	661,000	1,178
46	Mexico	4,740	2,667,725	563
47	Morocco	9	214	24
48	Myanmar	450	13,000	29
49	Netherlands	200	57,580	288
50	Neutral Zone	158	341,000	2,158
51	New Zealand	52	36,815	708
52	Nigeria	1,824	1,902,000	1,043
53	Norway	386	2,176,888	5,640
54	Oman	1,243	732,158	589
55	Pakistan	115	60,681	528

	Country	Producing Oil Wells	Average Daily Production BOPD	Average per Well, BOPD
56	PNG	23	52,380	2,277
57	Peru	3,157	116,992	37
58	Philippines	4	8,942	2,236
59	Poland	2,302	3,631	1.6
60	Qatar	238	425,000	1,786
61	Ras Al Khaima	7	1,000	142
62	Saudi Arabia	1,400	8,137,000	5,812
63	Serbia	646	22,000	34
64	Sharjah	31	40,000	1,290
65	South Africa	5	4,100	820
66	Spain	45	21,607	480
67	Suriname	223	5,000	22
68	Syria	963	518,000	538
69	Thailand	314	51,431	164
70	Trinidad & Tobago	3,262	135,415	42
71	Tunisia	181	106,157	587
72	Turkey	732	82,000	112
73	UK	735	1,928,731	2,624
74	USA	602,197	7,170,969	12
75	Venezuela	12,140	2,314,000	191
76	Vietnam	100	105,000	1,050
77	Yemen	179	176,000	983
78	Zaire	60	23,763	396
		925,749	60,040,391	65

From: *Oil & Gas Journal, Worldwide Production Report,* Vol. 91, No. 52 Dec. 1993

APPENDIX E

SELECTED US ENERGY STATISTICS (1992)

Stripper Oil Wells	462,823	Average Production 2.23 BOPD
(78% of total oil)		Remaining Recoverable
		7,500 bbls
		RLI = 9 years (1/RLI = 11%)
		17,500+ shut-in per year
Other Wells	149,956	Average Production 39 BOPD
		Remaining Recoverable
		142 MBBL/Well
Total Oil Wells	594,189	Average Production 12.1 BOPD
		(94% on artificial lift)
		Remaining 41.5 MBBLS/Well
Total Gas Wells	280,899	Average Production 182 MCFD
		Remaining Recoverable
		143.5 TCF
		Remaining Recoverable
		510 MMCF/Well

Total U.S. O&G Wells	875,088	1/1/93 currently producing
Cumulative U.S. Wells	3,182,606	1/1/93 producing and abandoned

Average Daily Oil Production	6,882 MBOPD
Average Daily Gas Production	47 BCFD

Seismic Crew Count	847
Active Drilling Rigs	721

Wells Drilled

New Field W/C	1,411	Average Depth 6,000 ft
Other Exploratory	1,851	
Oil	7,975	
Gas	6,425	
Dry	6,038	
Service Wells	863	
	————	
Total	21,301	
Success Ratio	70%	

Proved U.S. Reserves

Oil	24,682 MMBBLS
Gas	167 TCF
NGL	7,464 MMBBLS

Reserve Life Index

Oil	10 Years
Gas	10 Years

From: United States Department of Energy —
Energy Information Administration
American Petroleum Institute
Independent Petroleum Association of America
Baker Hughes Rig Count
Seismic Crew Count Society of Exploration Geophysicists

APPENDIX F

CONVERSION FACTORS

Btu Equivalents: Oil, Gas, Coal, and Electricity

One British Thermal Unit (Btu) is equal to the heat required to raise the temperature of one pound of water (approximately one pint) one degree Fahrenheit at or near its point of maximum density.

One barrel (42 gallons) of crude oil
- = 5,800,000 Btu of energy
- = 5,614 cubic feet of natural gas
- = 0.22 tons of bituminous coal
- = 1,700 kw hours of electricity

One cubic foot of natural gas (dry)
- = 1,032 Btu of energy
- = 0.000178 barrels of oil
- = 0.000040 tons of bituminous coal
- = 0.30 kw hours of electricity

One short ton (2,000 pounds) of bituminous coal
- = 26,200,000 Btu of energy
- = 5.42 barrels of oil
- = 25,314 cubic feet of natural gas
- = 7,679 kw hours of electricity

One kilowatt (kw) hour of electricity
- = 3,412 Btu of energy
- = 0.000588 barrels of oil
- = 3.306 cubic feet of natural gas
- = 0.00013 tons of bituminous coal

One gigajoule (10^9J) GJ

= 947,820 BtU of energy

GJ = gigajoules (10^9 joules) ≈ .9 MCF gas

TJ = terajoules (10^{12} joules) ≈ .9 MMCF gas

PJ = petajoules (10^{15} joules) ≈ .9 BCF gas

One British thermal unit (Btu)

= 1,055.06 Joules

METRIC CONVERSIONS

One metric ton of crude oil

= 2,204 pounds

= 7–7.5 barrels of oil

One cubic meter of natural gas

= 35.314 cubic feet

One cubic meter of liquid

= 6.2888 barrels

One liter of liquid

= 1.057 quarts

Distance

1 foot	= 0.305 meters	
1 meter	= 3.281 feet	
1 statute mile	= 1.609 kilometers	= 0.868 nautical miles
1 nautical mile	= 1.852 kilometers	= 1.1515 statute miles

Area

1 square mile	= 640 acres	= 2.59 square kilometers	= 259.0 square hectares
1 square kilometer	= 0.368 miles	= 100 hectares	= 247.1 acres
1 acre	= 43,560 square feet		= 0.405 hectares
1 hectare	= 2.471 acres		

Volume

1 cubic foot	= 0.028317 cubic meters
1 cubic meter	= 35.514667 cubic feet
1 cubic meter	= 6.2898 barrels
1 U.S. gallon	= 3.7854 liters
1 liter	= 0.2642 U.S. gallons
1 barrel	= 42 gallons = 158.99 liters

Weight

1 short ton	= 0.907185 metric tons = 0.892857 long tons
	= 2000 pounds
1 long ton	= 1.01605 metric tons = 1.120 short tons
	= 2240 pounds
1 metric ton	= 0.98421 long tons = 1.10231 short tons
	= 2204.6 pounds

APPENDIX G

METRIC U.S. CONVERSIONS

Unit	Metric Equivalent	U.S. Equivalent
acre	0.40468564 hectares	43,560 square feet
barrel	158.98729 liters	42 gallons
degrees Celsius	water: boils 100° freezes 0°	multiply by 1.8, add 32° = Fahrenheit
fathom	1.8288 meters	6 feet
hectare	10,000 square meters	2.471 acres
kilogram	0.001 metric tons	2.2046 pounds
kilometer	1,000 meters	0.62137 miles
square kilometer	100 hectares	247.1 acres
nautical mile	1.852 kilometers	1.151 statute miles
liter	0.001 cubic meters	61.02374 cubic inches 1.0567 quarts, liquid
meter	100 centimeters	3.280839 feet
ton, long or deadweight	1,016.047 kilograms	2,240 pounds
ton, metric	1,000 kilograms	2,204.623 pounds
ton, short	907.18474 kilograms	2,000 pounds

APPENDIX H

BARRELS PER METRIC TON VS. API GRAVITY

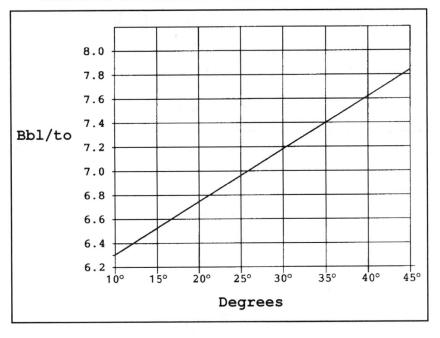

APPENDIX I

RELATIVE OIL PRICE VS. API GRAVITY

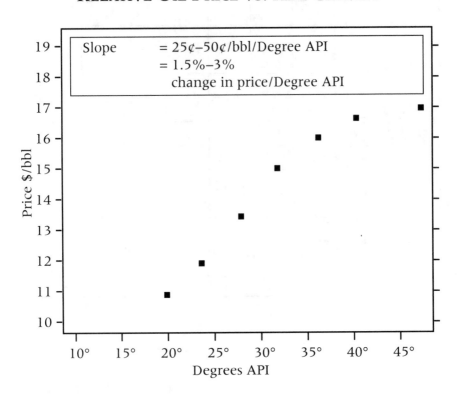

Depending upon market conditions and the particular region, the slope of the curve can change dramatically.

Percent Sulfur

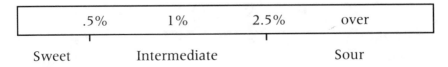

The average sulfur content for U.S. refineries is 1%.

APPENDIX J

NATURAL GAS PRODUCTS

HYDROCARBON SERIES					TERMINOLOGY
C1	C2	C3	C4	C5+	
◄—LNG --► |					Liquified Natural Gas
◄—CNG --► |					Compressed Natural Gas
		|◄—LPG —► |			Liquified Petroleum Gas
	|◄-------NGL —————————►				Natural Gas Liquids
				|COND|	Condensate

Methane   Ethane   Propanes   Butanes   Pentanes+

Physical Constants – Light Hydrocarbons

	Degrees API		MCF/BBL	#/BBL	BBLS/Ton Metric	MCF/Ton Metric Ton	MCF/Ton Long Ton	MCF/Ton Short Ton
Methane	340.2	C1	2.483	104.9	21.0	52.17	53.02	47.34
Ethane	253.1	C2	1.647	130.5	16.9	27.82	28.27	25.24
Propane	147.2	C3	1.526	177.3	12.4	18.97	19.28	17.21
I–Butane	119.3							
N–Butane	110.6	C4	1.304	200.4	11.0	14.34	14.58	13.01
I–Pentane	95.0							
N–Pentane	92.7	C5	1.154	219.5	10.0	11.59	11.78	10.51
N–Hexane	81.6	C6	1.022	232.2	9.5	9.70	9.86	8.80
N–Heptane	74.1	C7	0.915	241.6	9.1	8.35	8.48	7.57
N–Octane	68.7	C8	0.821	247.1	8.9	7.32	7.44	6.65
N–Nonane	64.5	C9	0.747	252.6	8.7	6.52	6.62	5.91
Pentanes +		C5+	0.932	238.6	9.2	8.61	8.75	7.81
Hexanes +		C6+	0.876	243.4	9.1	7.93	8.06	7.20
Water				349.8	6.3			

Metric Ton = 2,204 #
Long Ton = 2,240 #
Short Ton = 2,000 #

Water   8.328   Pounds/US Gallon
      623   Pounds/Cubic Meter

APPENDIX K

H$_2$S AND NATURAL GAS

1 ppm	= .0001%	Detectable by odor

10 ppm = .001% 7 hours exposure allowable

over 20 ppm Protective equipment is necessary

100 ppm = .01% Kills sense of smell in 3–15 minutes, may burn eyes and throat

200 ppm = .02% Kills smell rapidly, burns eyes and throat

500 ppm = .05% Sense of reasoning and balance impaired
Respiratory problems within 2–15 minutes
Artificial resuscitation needed promptly

700 ppm = .07% Unconsciousness occurs quickly. Breathing stops, and death results if not immediately resuscitated.

1,000 ppm = .10% Immediate unconsciousness

APPENDIX L

PRESENT VALUE OF ONE TIME PAYMENT

Present Value of $1 = $\dfrac{1}{(1+i)^{(n-.5)}}$

(Midyear discounting)

Period (n)	5%	10%	15%	20%	25%	30%	35%
1	.976	.953	.933	.913	.894	.877	.861
2	.929	.867	.811	.761	.716	.765	.638
3	.885	.788	.705	.634	.572	.519	.472
4	.843	.717	.613	.528	.458	.339	.350
5	.803	.651	.533	.440	.366	.307	.259
6	.765	.592	.464	.367	.293	.236	.192
7	.728	.538	.403	.306	.234	.182	.142
8	.694	.489	.351	.255	.188	.140	.105
9	.661	.445	.305	.212	.150	.108	.078
10	.629	.404	.265	.177	.120	.083	.058
11	.599	.368	.231	.147	.096	.064	.043
12	.571	.334	.200	.123	.077	.049	.032
13	.543	.304	.174	.102	.061	.038	.023
14	.518	.276	.152	.085	.049	.029	.017
15	.493	.251	.132	.071	.039	.022	.013
16	.469	.228	.115	.059	.031	.017	.010
17	.447	.208	.100	.049	.025	.013	.007
18	.426	.189	.087	.041	.020	.010	.005
19	.406	.171	.075	.034	.016	.008	.004
20	.386	.156	.066	.029	.013	.006	.003
21	.368	.142	.057	.024	.010	.005	.002
22	.350	.129	.050	.020	.008	.004	.002
23	.334	.117	.043	.017	.007	.003	.001
24	.318	.106	.037	.014	.005	.002	.001
25	.303	.097	.033	.011	.004	.002	.001

APPENDIX M

REFERENCES AND SOURCES OF INFORMATION

Barrows, G. *Worldwide Concession Contracts and Petroleum Legislation.* Tulsa: PennWell Publishing Company, 1983.

Barry, R. *The Management of International Oil Operations.* Tulsa: PennWell Publishing Company, 1993.

Beck, R. *Oil Industry Outlook Ninth Edition 1993–1997 Projection to 2001.* Tulsa: PennWell Publishing Company, 1992.

Bosson, R., and M. Varon, *The Mining Industry and the Developing Countries.* Washington, D.C.: World Bank, 1977.

Boulos, A. "How a Domestic Oil Company Goes International: A Strategy for Success." Production Sharing Contracts Conference Proceedings, AIC Conferences, Houston, March 1994.

Burke, F., and R. Dole, *Business Aspects of Petroleum Exploration in Non-Traditional Areas.* BMC, 1991.

Darden, M. *Legal Research Checklist for International Petroleum Operations,* Section on Natural Resources, Energy, and Environmental Law. Monograph Series Number 20, American Bar Association, 1994.

_____. "Lifting Agreements in International Petroleum Operations." Presentation to the Energy Law Section of the Dallas Bar Association, 16 February 1994.

Derman, A. *International Oil and Gas Joint Ventures: A Discussion with Associated Form Agreements,* Section on Natural Resources, Energy, and Environmental Law. Monograph Series Number 16, The American Bar Association, 1992.

Dur, S. "Negotiating PSC Terms." Production Sharing Contracts Conference Proceedings, AIC Conferences, Houston, March 1994.

Fee, D. *Oil & Gas Databook for Developing Countries,* 2nd Edition. Commission of the European Communities, 1988.

Foley, M. and G. Gussis, "Tax & Fiscal Regimes: A Comparative Analysis." Oil & Gas Production-Sharing Contracts (PSCs), Concessions and New Petroleum Ventures in the Asia-Pacific Basin, Conference Proceedings, Institute for International Research, Houston, April 1993.

Frame, S. "Risk & Reward—Assessing Upstream Potential." Oil & Gas Production-Sharing Contracts (PSCs), Concessions and New Petroleum Ventures in the Asia-Pacific Basin, Conference Proceedings, Institute for International Research, Houston, April 1993.

Gosain, V. "How to Identify Creditable Foreign Taxes." *U.S. Taxation of International Operations,* Warren Gorham Lamont, 1992, pp. 5,471–5,495.

Hallmark, T. "Political Risk: Assessing dangers in international exploration, development." *Offshore Magazine,* (May 1991): 27–34.

Hook, F. "Sovereign Risk and the Resource Industries." *PetroMin Magazine,* (April 1992): 39–45.

Hossain, K. *Law and Policy in Petroleum Development* London,: Frances Pinter (Publishers) Ltd., 1979.

Johnston, D. "Current Status of Petroleum Fiscal Systems in the World Oil & Gas Industry." Production Sharing Contracts Conference Proceedings, AIC Conferences, Houston, March 1994.

Johnston, D. "International Petroleum Fiscal Systems: Production Sharing Contracts." *Petroleum Accounting and Financial Management Journal*, Special International Issue, (Summer 1994).

Johnston, D. "The Production Sharing Concept." *PetroMin Magazine*—Singapore, (August 1992): 26–34.

Khong Cho Oon. *The politics of oil in Indonesia: Foreign company-host government relations*. Cambridge: Cambridge University Press, 1986.

McPherson, C. and K. Palmer, "New Approaches to Profit Sharing in Developing Countries." *Oil & Gas Journal*, (25 June 1984): 119–128.

Mikdashi, Z. *The International Politics of Natural Resources*. Ithaca and London: Cornell University Press, 1976.

Mikesell, R. *Petroleum Company Operations and Agreements in the Developing Countries*. Washington, D.C.: Resources for the Future, Inc.,1984. Distributed by Johns Hopkins University Press.

Mustafaoglu, M. "Comparison of Major International Petroleum Tax Systems." *Journal of Petroleum Technology, (*Society of Petroleum Engineers) *(*October 1981): 1835–1843.

Ricardo, D. *The Principles of Political Economy and Taxation*. London: Dent & Sons Ltd. Printed in Great Britain by Biddles Ltd, Guildford, Surrey, First published in 1911. Reprinted in 1976.

Smith, D., "Comparison of Fiscal Terms in the Far East, South America, North Africa and C.I.S." Oil & Gas Production-Sharing Contracts (PSCs), Concessions and New Petroleum Ventures in the Asia-Pacific Basin, Conference Proceedings, Institute for International Research, Houston, April 1993.

Smith, D. and L. Wells, *Negotiating Third-World Mineral Agreements*. Cambridge, Mass:, Ballinger Publishing Company, 1975.

Stauffer, T. and J. Gault, "Effects of Petroleum Tax Design upon Exploration and Development." SPE Paper 9576, from Proceedings of the 1981 Hydrocarbon Economics and Evaluations Symposium, pp. 199–211.

Stauffer, T. "Political Risk and Overseas Investment" SPE Paper 18514, from Proceedings of the SPE Symposium on Energy, Finance and Taxation Policies held in Washington, D.C., September 1988.

Wood, D. "Appraisal of economic performance of global exploration contracts." Two-part series, *Oil & Gas Journal,* (October 29, 1990; November 5, 1990): 48–53, 50–53.

_____. *Legal Framework for the Treatment of Foreign Investment*. Volume I, Washington, D.C.: The World Bank, 1992.

_____. *Legal Framework for the Treatment of Foreign Investment*. Volume II, Washington, D.C.: The World Bank, 1992.

_____. *Petroleum Accounting and Financial Management Journal,* Special International Issue. Summer 1994, Denton, Texas: Institute of Petroleum Accounting, (Johnston, D. ed., 1994)

_____. *The Petroleum Report, Indonesia's Petroleum Sector.* Jakarta, Indonesia: Embassy of the United States of America, June 1990.

_____. *Studies in Energy Tax Policy.* Edited by G. Brannon. Cambridge, Mass.: Ballinger Publishing Company, 1975.

GLOSSARY

Abrogate. To officially abolish or repeal a treaty or contract through legislative authority or an authoritative act.

Accelerated depreciation. Writing off an asset through depreciation or amortization at a rate that is faster than normal accounting straight line depreciation. There are a number of methods of accelerated depreciation but they are usually characterized by higher rates of depreciation in the early years than the latter years in the life of the asset. Accelerated depreciation allows for lower tax rates in the early years.

Ad Valorem. Latin for *according to value*, a tax on goods or property, based upon value rather than quantity or size.

Affiliate. Two companies are affiliated when one owns less than a majority of the voting stock of the other or when they are both subsidiaries of a third parent company (see Subsidiary). A subsidiary is an affiliate of its parent company.

Amortization. An accounting convention designed to emulate the cost or expense associated with reduction in value of an intangible asset (see Depreciation) over a period of time. Amortization is a noncash expense. Similar to depreciation of tangible capital costs, there are several techniques for amortization of intangible capital costs:
 • Straight Line Decline (SLD)
 • Double Declining Balance (DDB)

- Declining Balance (DB)
- Sum of Year Digits (SYD)
- Unit of Production

Arbitration. A process in which parties to a dispute agree to settle their differences by submitting their dispute to an independent individual or group for settlement. Each side of the dispute chooses an arbitrator, and those two choose a third. The third arbitrator acts as the chairman of the tribunal which then hears and reviews both sides of the dispute. The tribunal then renders a decision that is final and binding.

Book Value. (1) The value of the equity of a company. Book value per share is equal to the equity divided by the number of shares of common stock. Fully diluted book value is equal to the equity less any amount that preferred shareholders are entitled to divided by the number of shares of common stock. (2) Book value of an asset or group of assets is equal to the initial cost less DD&A.

Branch. An extension of a parent company, but not a separate independent entity. Subsidiary companies are normally taxed as profits are distributed as opposed to branch profits which are taxed as they accrue.

Calvo Clause. A relatively obsolete contract clause once promoted in Latin American countries where the contractor explicitly renounced the protection of its home government over its operation of the contract. The objective of the Calvo Doctrine was to direct disputes to local jurisdictions and avoid international arbitration.

Capitalization. All money invested in a company including long term debt (bonds), equity capital (common and preferred stock), retained earnings, and other surplus funds.

Capitalization Rate. The rate of interest used to convert a series of future payments into a single present value.

Capitalize. (1) In an accounting sense, the periodic expensing (amortization) of capital costs through depreciation or depletion. (2) To convert an (anticipated) income stream to a present value by dividing by an interest rate, as in the dividend discount model. (3) To record capital outlays as additions to asset value rather than as expenses.

Generally, expenditures that will yield benefits to future operations beyond the accounting period in which they are incurred are capitalized—that is, they are depreciated at either a statutory rate or a rate consistent with the useful life of the asset.

Cash Flow. (1) Net income plus depreciation, depletion, and amortization and other noncash expenses. Usually synonymous with cash earnings and operating cash flow. (2) An analysis of all the changes that affect the cash account during an accounting period.

Central Bank. The primary government-owned banking institution of a country. The central bank usually regulates all aspects of foreign exchange in and out of the country. It actively intervenes in the acquisition and sale of its own currency in foreign exchange markets primarily to maintain stability in the value of the country's currency.

CIF. Cost insurance and freight is included in the contract price for a commodity. The seller fulfills his obligations when he delivers the merchandise to the shipper, pays the freight and insurance to the point of (buyer's) destination, and sends the buyer the bill of lading, insurance policy, invoice, and receipt for payment of freight. The following example illustrates the difference between an FOB Jakarta price and a CIF Yokohama price for a ton of LNG (see FOB).

FOB Jakarta	$170/ton	also called *netback price*
	+ 30/ton	Freight charge
CIF Yokohama	$200/ton	

Commercial Discovery. In popular usage, the term applies to any discovery that would be economically feasible to develop under a given fiscal system. As a contractual term, it often applies to the requirement on the part of the contractor to demonstrate to the government that a discovery would be sufficiently profitable to develop from both the contractor's and government's points of view. A field that satisfied these conditions would then be granted commercial status, and the contractor would then have the right to develop the field.

Concession. An agreement between a government and a company that grants the company the right to explore for, develop, produce, transport, and market hydrocarbons or minerals within a fixed area for a specific amount of time. The concession and production and sale of hydrocarbons from the concession is then subject to rentals, royalties, bonuses, and taxes. Under a concessionary agreement the company would hold title to the resources that are produced.

Consortium. A term that applies to a group of companies operating jointly, usually in a partnership with one company as operator in a given permit, license, contract area, block, etc.

Contractor. An oil company operating in a country under a production sharing contract or a service contract on behalf of the host government for which it receives either a share of production or a fee.

Contractor After-Tax Equity Split. Same as contractor take.

Contractor Take. The total contractor after-tax share of profits.

Cost Insurance and Freight (CIF). A transportation term that reflects the price of a commodity at the point of sale which includes all transportation costs including insurance, etc. (see CIF).

Cost of Capital. The minimum rate of return on capital required to compensate debt holders and equity investors for bearing risk. Cost of capital is computed by weighting the after-tax cost of debt and equity according to their relative proportions in the corporate capital structure.

Cost Oil. A term most commonly applied to production sharing contracts which refers to the oil (or revenues) used to reimburse the contractor for exploration costs, development capital costs, and operating costs.

Country Risk. The risks and uncertainties of doing business in a foreign country, including political and commercial risks (see Sovereign Risk).

Creeping Nationalization or **Creeping Expropriation.** A subtle means of expropriation through expanding taxes, restrictive labor legislation, labor strikes, withholding work permits, import restrictions, price controls, and tariff policies. The difference between nationalization and expropriation is that nationalization is usually on an industry-wide level and expropriation focuses on a particular company.

Debt Service. Cash required in a given period, usually one year, for payments of interest and current maturities of principal on outstanding debt. In corporate bond issues, the annual interest plus annual sinking fund payments.

Depletion. (1) Economic depletion is the reduction in value of a wasting asset by the removal of minerals. (2) Depletion for tax purposes (depletion allowance) deals with the reduction of mineral resources due to removal by production or mining from an oil or gas reservoir or a mineral deposit.

Depreciation. An accounting convention designed to emulate the cost or expense associated with reduction in value of an asset due to wear and tear, deterioration, or obsolescence over a period of time. Depreciation is a noncash expense. There are several techniques for depreciation of capital costs:
• Straight Line Decline
• Double Declining Balance
• Declining Balance
• Sum of Year Digits
• Unit of Production (see Unit of Production)

Dilution Clause. In a joint operating agreement, a clause that outlines a formula for the dilution of interest of a working-interest partner if that partner defaults on a financial obligation. Also called a withering clause.

Direct Tax. A tax that is levied on corporations or individuals— the opposite of an indirect tax, such as a value-added tax (VAT) or sales taxes.

Disposal. This term usually refers to transportation and sales of crude or gas from the field.

Dividend Withholding Tax. A tax levied on dividends or repatriation of profits. Tax treaties normally try to reduce these taxes whether they are so named or simply operate in the same manner as a withholding tax.

Dollars-of-the-Day. A term usually associated with cost esti-mates that indicate the effects of anticipated inflation have been taken into account. For example, if a well costs $5 million right now in "today's dollars" (the opposite of dollars-of-the-day), then the cost of the well two years from now might be estimated at $5.51 million in dollars-of-the-day assuming a 5% inflation fac-tor. Also called *escalated dollars.*

Domestication. A form of creeping nationalization where host government enacts legislation that forces foreign-owned enter-prises to surrender various degrees of ownership and/or control to nationals.

Double Taxation. (1) In economics a situation where income flow is subjected to more than one tier of taxation under the same domestic tax system—such as state/provincial taxes, then federal taxes, or federal income taxes and then dividend taxes. (2) International double taxation is where profit is taxed under the system of more than one country. It arises when a taxpayer or taxpaying entity resident (for tax purposes) in one country gener-ates income in another country. It can also occur when a taxpay-ing entity is resident for tax purposes in more than one country (see Dual Residence).

Double Taxation Treaty. Formal agreement between countries to reduce or eliminate double taxation. A bilateral tax treaty is a treaty between two countries to coordinate taxation provisions which would otherwise create double taxation. A multilateral tax treaty involves three or more countries for the same purpose. The United States has few treaties with oil-producing nations.

Dual Residence. When a taxpaying entity is resident for tax purposes in more than one country. This can happen when differ-ent countries apply the tests for determining residence and the company passes the test in more than one country.

Dutch Disease. The adverse results of large-scale positive shock to a single sector of a nation's economy, so named because of the problems associated with large-scale development of the Groningen Gas field in the Netherlands in the 1970s. Typically the sector of economy that is booming causes widespread inflation and other sectors, particularly agriculture, suffer from inability to attract workers. The drastic increase in foreign exchange can cause problems with local currencies and fiscal and monetary problems can occur without proper management.

Economic Rent. The difference between the value of production and the cost to extract it. The extraction cost consists of normal exploration, development, and operating costs as well as a sufficient share of profits for the industry. Economic rent is what the governments try to extract as efficiently as possible.

Entitlements. The shares of production to which the operating company, the working-interest partners, and the government or government agencies are authorized to lift. Entitlements are based upon royalties, cost recovery, production sharing, taxation, working-interest percentages, etc. Generally, legal entitlement equals Profit Oil plus Cost Oil in a PSC.

Equity Oil. Usually this term refers to oil or revenues after cost recovery (or cost oil). It is also referred to as profit oil or share oil—terms that are most often associated with PSCs. Generally speaking, the analog to equity oil in a concessionary system would be pretax cash flow. Like pretax cash flow, equity oil may also be subject to taxation.

Eurodollar. U.S. currency that is held in foreign banks (mostly European) that is used in settling international transactions.

Excise Tax. A tax based either on production, sale, or consumption of a specific commodity such as tobacco, coffee, gasoline, or oil.

Exclusion of Areas. see Relinquishment

Expense. (1) In a financial sense, a noncapital cost associated most often with operations or production. (2) In accounting, costs incurred in a given accounting period as expenses and charged against revenues. To expense a particular cost is to charge it against income during the accounting period in which it was spent. The opposite would be to capitalize the cost and charge it off through some depreciation schedule.

Exploratory Well. A well drilled in an unproved area. This can include: (1) a well in a proved area seeking a new reservoir in a significantly deeper horizon, (2) a well drilled substantially beyond the limits of existing production. Exploratory wells are defined partly by distance from proved production and by degree of risk associated with the drilling. Wildcat wells involve a higher degree of risk than exploratory wells.

Expropriation. Similar to the concept of nationalization or outright seizure or confiscation of foreign assets by a host government. With expropriation the confiscation is directed toward a particular company; nationalization is where a government confiscates a whole industry. Expropriation is legal but theoretically must be accompanied by prompt, adequate, and effective compensation and must be in the public interest.

Fair Market Value (FMV) of Reserves. Often defined as a specific fraction of the present value of future net cash flow discounted at a specific discount rate. The most common usage defines FMV at $^2/_3$–$^3/_4$ of the present value of future net cash flow discounted at the prime interest rate plus .75–1 percentage point.

Finding Cost. The amount of money spent per unit (barrel of oil or MCF of gas) to acquire reserves. Includes discoveries, acquisitions, and revisions to previous reserve estimates.

Fiscal Marksmanship. The ability of authorities to predict with any degree of accuracy or certainty the tax revenues that may fall due to be paid to the government. In the petroleum industry, it is particularly difficult to estimate accurately what revenues may be generated for countries with little or no exploration history.

Fiscal System. Technically, the legislated taxation structure for a country including royalty payments. In popular language, the term includes all aspects of contractual and fiscal elements that make up a given government-foreign oil company relationship.

FOB. Free on Board. A transportation term that means that the invoice price includes transportation charges to a specific destination. Title is usually transferred to the buyer at the FOB point by way of a bill of lading. For example, FOB New York means the buyer must pay all transportation costs from New York to the buyer's receiving point. FOB plus transportation costs equals CIF price (see CIF—Cost Insurance Freight).

Foreign Corrupt Practices Act. Sometimes referred to as antib-ribery legislation. It is illegal for a U.S. company or individual to knowingly pay a foreign official in order to obtain or to retain business. This includes commissions or payments to agents or intermediaries with the knowledge that all or a part of the payments will be given to a foreign official. The FCPA also has various record-keeping and reporting requirements.

Foreign Tax Credit. Taxes paid by a company in a foreign country may sometimes be treated as taxes paid in the company's home country. These are creditable against taxes and represent a direct dollar-for-dollar reduction in tax liability. This usually applies to foreign income taxes paid and credited against home-country income taxes. Other taxes which may not qualify for a tax credit may nevertheless qualify as deductions against home-country income tax calculation.

Franked Dividends. Dividends that have already been taxed at the corporate level and are therefore either not subject to withholding tax or the taxes paid are creditable against withholding taxes.

Gazette. To announce officially license round offering or results, or publication of notification of acceptance of bids in official government publication (gazette). To *gazette* means to offer block, as in "The licenses have not been gazetted yet."

Gold Plating. When a company or contractor makes unreasonably large expenditures due to lack of cost-cutting incentives. This kind of behavior could be encouraged where a contractor's compensation is based in part on the level of capital and operating expenditure.

Government After-Tax Equity Split. Same as government take.

Government Take. The total government share of profit oil or revenues not associated with cost recovery. Same as government after-tax equity split and government marginal take.

Hard Currency. Currency in which there is widespread confidence and a broad market such as that for the U.S. dollar, the British pound, Swiss francs, or Japanese yen. The opposite would be soft currency where there is a thin market and the currency fluctuates erratically in value. Soft currencies usually are based upon unrealistic exchange rates that do not reflect the market value of the currency.

Hectare. Metric unit of area equal to 10,000 m_2 or 100 *ares*, which also equals 2.471 acres.

Hull Formula. Compensation for expropriation, in the language of many bilateral and multilateral investment treaties, should be "prompt, adequate, and effective." This is known as the Hull Formula. Alternate wording found in other treaties includes, "fair and equitable,""reasonable,""market value at date of expropriation," etc.

Hurdle Rate. Term used in investment analysis or capital budgeting that means the required rate of return in a discounted cash flow analysis. Projects to be considered viable must at least *meet the hurdle rate.* Investment theory dictates that the hurdle rate should be equal to or greater than the incremental cost of capital.

Hydrocarbon Series. The various components of crude oil and natural gas composed of carbon and hydrogen atoms.

Hydrocarbon Series

C1	- Methane	- CH_4
C2	- Ethane	- C_2H_6
C3	- Propanes	- C_3H_8
C4	- Butanes	- C_4H_{10}
C5	- Pentanes	- C_5H_{12}
C6	- Hexanes	- C_6H_{14}
C7	- Heptanes	- C_7H_{16}
C8	- Octanes	- C_8H_{18}
C9	- Nonanes	- C_9H_{20}
C10 -	Decanes	- $C_{10}H_{22}$
and so forth		

Incentives. Fiscal or contractual elements emplaced by host governments that make petroleum exploration or development more economically attractive. Includes such things as:
- Royalty holidays
- Tax holidays

- Tax credits
- Reduced government participation
- Lower government take
- Investment credits/uplifts
- Accelerated depreciation

Inconvertibility. Inability of a foreign contractor to convert payments received in soft local currency into home country or hard currency such as dollars, pounds, or yen.

Indirect Tax. A tax that is levied on consumption rather than income (see Direct Tax). Examples of indirect taxes include value-added taxes, sales taxes, or excise taxes on luxury items.

Intangible Drilling and Development Costs (IDCs). Expenditures for wages, transportation, fuel, fungible supplies used in drilling and equipping wells for production.

Intangibles. All intangible assets such as goodwill, patents, trademarks, unamortized debt discounts, and deferred charges.

Investment Credit. A fiscal incentive where the government allows a company to recover an additional percentage of tangible capital expenditure. For example, if a contractor spent $10 million on expenditures eligible for a 20% investment credit, then the contractor would actually be able to recover $12 million through cost recovery (see Uplift). These incentives can be taxable. Sometimes the investment credit is mistakenly referred to as an investment tax credit.

Joint Venture. The term applies to a number of partnership arrangements between individual oil companies or between a company and a host government. Typically an oil company or consortium (contractor group) carries out sole risk exploration efforts with a right to develop any discoveries made. Development and production costs then are prorated between the partners, which may include the government.

Letter of Credit. An instrument or document from a bank to another party indicating that a credit has been opened in that party's favor guaranteeing payment under certain contractual conditions. The conditions are based upon a contract between the two parties. Sometimes called a *performance letter of credit* which is issued to guarantee performance under the contract.

Letter of Intent. A formal letter of agreement signed by all parties to negotiations after negotiations have been completed outlining the basic features of the agreement, but preliminary to formal contract signing.

License. An arrangement between an oil company and a host government regarding a specific geographical area and petroleum operations. In more precise usage, the term may apply to the development phase of a contract after a commercial discovery has been made (see Permit).

Lifting. The amount of crude oil an operator produces and sells, or the amount each working-interest partner (or the government) takes. The liftings may actually be more or less than actual entitlements which are based on royalties, working-interest percentages, and a number of other factors. If an operator or partner has taken and sold more oil than it was actually entitled to, then it is in an *overlifted* position. Conversely if a partner has not taken as much as it was entitled to it is in an *underlifted* position (see Nomination and Entitlements).

Limitada. Business entity which resembles a partnership with liability of all members limited to their contribution and no general partner with unlimited liability. Normally treated as a partnership by the United States for tax purposes. Similar to a limited liability company in the United States, although the limitada was the forerunner.

London Interbank Offered Rate (LIBOR). The rate that the most creditworthy international banks that deal in Eurodollars will charge each other. Thus LIBOR is sometimes referred to as the *Eurodollar Rate*. International lending is often based on LIBOR rates. For example, a country may have a loan with interest pegged at *LIBOR plus 1.5%* (see Eurodollar).

Marginal Government Take. Same as government take.

Maximum Cash Impairment. Maximum negative cash balance in a cash flow projection.

Nationalization. Government confiscation of the assets held by foreign companies throughout an entire industry (see Expropriation).

Netback. Many royalty calculations are based upon gross revenues from some point of valuation, usually the last valve off of a production platform or at the boundary of a field or license area. The point of sale, however, may be different than the point of valuation. The statutory royalty calculation may allow the transportation costs from the point of valuation to the point of sale to be deducted. This is called a *netback formula*.

Nomination. Under a lifting agreement the amount of crude oil a working-interest owner is expected to lift. Each working-interest partner has a specific entitlement depending upon the level of production, royalties, their working interest, and their relative position (i.e., underlifted or overlifted), etc. Each working-interest partner must notify the operator (nominate) the amount of its entitlement that it will lift. Sometimes, depending upon the lifting agreement, the nomination may be more or less than the actual entitlement (see Liftings and Entitlements).

OPEC. Organization of Petroleum Exporting Countries founded in 1960 to coordinate petroleum prices of the members. Members include:
- Iran
- Iraq
- Kuwait
- Saudi Arabia
- Venezuela
- Qatar
- Indonesia
- Libya
- Abu Dhabi (UAE)
- Algeria
- Nigeria
- Ecuador (dropped out)
- Gabon

Operating Profit (or loss). The difference between business revenues and the associated costs and expenses exclusive of interest or other financing expenses, and extraordinary items, or ancillary activities. Synonymous with net operating profit (or loss), operating income (or loss), and net operating income (or loss).

OPIC. (Overseas Private Investment Corporation) A U.S. government agency founded under the Foreign Assistance Act of 1969 to administer the national investment guarantee program for investment in less developed countries (LDCs) through the issuance of insurance for risks associated with war, expropriation, and inconvertibility of payments in local currency.

Overlifting. Over/underlifting is the difference between actual contractor lifting during an accounting period and the contractor entitlements based upon cost recovery and profit oil, in the case of a PSC. A lifting is the actual physical volume of crude oil taken and sold.

Overspill. In international taxation, a situation where a taxpaying company has a credit for foreign taxes that is greater than its corporate tax liability in its home country, giving it has an unused and/or unusable tax credit.

Permit. In a loose sense the term is used to describe any arrangement between a foreign contractor and a host government regarding a specific geographical area and petroleum operations. In a more precise usage, the term may apply to the exploration phase of a contract before a commercial discovery has been made (see License).

Posted Price. The official government selling price of crude oil. Posted prices may or may not reflect actual market values or market prices.

Present Value. The value now of a future payment or stream of payments based on a specified discount rate.

Prime Lending Rate. The interest rate on short-term loans that banks charge to their most stable and creditworthy customers. The prime rate charged by major lending institutions is closely watched and is considered a benchmark by which other loans are based. For example, a less well-established company may borrow at prime plus 1%.

Production Sharing Agreement. This (PSA) is the same as a Production Sharing Contract (PSC). While at one time this term was quite common, it is used less frequently now, and the term *Production Sharing Contract* is becoming more common.

Production Sharing Contract. A contractual agreement between a contractor and a host government whereby the contractor bears all exploration costs and risks and development and production costs in return for a stipulated share of the production resulting from this effort.

Pro Forma. Latin for *as a matter of form*. A financial projection based upon assumptions and possible events that have not occurred. For example, a financial analyst may create a consolidated balance sheet of two nonrelated companies to see what the combination would look like if the companies had merged. Often a cash flow projection for discounted cash flow analysis is referred to as a pro forma cash flow.

Progressive Taxation. Where tax rates increase as the basis to which the applied tax increases. Or where tax rates decrease as the basis decreases. The opposite of regressive taxation.

Protocol. (1) Culturally dictated forms of ceremony and etiquette that govern business relationships, meetings, and negotiations. (2) Formal document primarily used in republics of the former Soviet Union signed by parties who attend meetings or negotiations indicating various minor agreements or stages of agreement reached. This document is not the same as a letter of intent, which is more formal and usually signifies that negotiations have been concluded.

Rate of Return Contract. Sometimes referred to as a Resource Rent Royalty. The government collects a share of cash flows in excess of the return required to generate investment. The government share is calculated by accumulating negative net cash flows at a specific threshold rate of return, and once the accumulated value becomes positive, the government takes a specified share.

Regressive Tax. Where tax rates become lower as the basis to which the applied tax increases. Or where tax rates increase as the basis decreases. The opposite of progressive taxation.

Reinvestment obligations. A fiscal term that requires the contractor/operator to set aside a specified percentage of profit oil or income after-tax that must be spent on domestic projects such as exploration.

Resource Rent Tax. Some economists refer to additional profits taxes as a resource rent tax. Australia has a specific tax based upon profits which is referred to as resource rent tax (RRT) (see Rate of Return Contract). Normally the RRT is levied after the contractor or oil company has recouped all capital costs plus a specified return on capital that supposedly will yield a fair return on investment.

Relinquishment A contract clause that refers to how much contract or license area a contractor must surrender or give back to the government during or after the exploration phase of a contract. Licenses are usually granted on the basis of an initial term with specific provisions for the timing and amount of relinquishment prior to entering the next phase of the contract. Also referred to as *exclusion of areas*.

Ringfencing. A cost center based fiscal device that forces contractors or concessionaires to restrict all cost recovery and or deductions associated with a given license (or sometimes a given field) to that particular cost center. The cost centers may be individual licenses or on a field-by-field basis.

For example, exploration expenses in one nonproducing block could not be deducted against income for tax calculations in another block. Under a PSC, ringfencing acts in the same way: cost incurred in one ring fenced block cannot be recovered from another block outside the ringfence.

Royalty Holiday. A form of fiscal incentive to encourage investment and particularly marginal field development. A specified period of time in years or months, during which royalties are not payable to the government. After the holiday period, the standard royalty rates are applicable (see Tax Holiday).

Seismic Option. A contractual arrangement or agreement between a host government and a contractor that gives the contractor a period during which it has exclusive rights over a geographic area or contract area during which the contractor will shoot additional seismic data. After data acquisition, processing, and interpretation, the contractor has the right to enter into an additional phase of the agreement or a more formal contract with the government for the area, which usually includes a drilling commitment.

Severence Tax. A tax on the removal of minerals or petroleum from the ground, usually levied as a percentage of the gross value of the minerals removed. The tax can also be levied on the basis of so many cents per barrel or per million cubic feet of gas.

Shelf Company. An incorporated entity which has no assets and or income but has gone through the process of registration and licensing. Some operations in foreign countries are started with acquisition of a shelf company because of the long delays that can be experienced in incorporating a company.

Sinking Fund. Money accumulated on a regular basis in a separate account for the purpose of paying off an obligation or debt.

Sliding Scales. A mechanism in a fiscal system that increases effective taxes and/or royalties based upon profitability or some proxy for profitability, such as increased levels of oil or gas production. Ordinarily each tranche of production is subject to a specific rate, and the term *incremental sliding scale* is sometimes used to further identify this.

Sovereign Risk. Also called country risk or political risk—refers to the risks of doing business in a foreign country where the government may not honor its obligations or may default on commitments. Encompasses a variety of possibilities including nationalization, confiscation, expropriation, among others (see Country Risk).

Spot Market. Commodities market where oil (or other commodities) are sold for cash and the buyer takes physical delivery immediately. Futures trades for the current month are also called *spot market trades*. The spot market is mostly an over-the-counter market conducted by telephone and not on the floor of an organized commodity exchange.

Spot Price. The delivery price of a commodity traded on the spot market. Also called the *cash price*.

Subsidiary. A company legally separated from but controlled by a parent company who owns more than 50% of the voting shares. A subsidiary is always by definition an affiliate company (see Affiliate). Subsidiary companies are normally taxed as profits are distributed as opposed to branch profits, which are taxed as they accrue.

Sunk Costs. There are a number of categories of sunk costs:
- Tax Loss Carry Forward (TLCF)
- Unrecovered Depreciation Balance
- Unrecovered Amortization Balance
- Cost Recovery Carry Forward

These costs represent previously incurred costs that will ultimately flow through cost recovery or will be available as deductions against various taxes (if eligible).

Surrender. (see Relinquishment). Surrender is most often used synonymously with relinquishment in the context of area reduction. However, the term also is used to describe a contractor's option to withdraw from a license or contract at or after various stages in a contract.

Take-or-Pay Contract. A type of contract where specific quantities of gas (usually daily or annual rates) must be paid for, even if delivery is not taken. The purchaser may have the right in following years to take gas that had been paid for but not taken.

Tax. A compulsory payment pursuant to the authority of a government. Fines, penalties, interest, and customs duties are not taxes.

Tax Haven. A country where certain taxes are low or nonexistent in order to increase commercial and financial activity.

Tax Holiday. A form of fiscal incentive to encourage investment. A specified period of time, in years or months, during which income taxes are not payable to the government. After the holiday period, the standard tax rates apply.

Tax Treaty. A treaty between two (bilateral) or more (multilateral) nations which lowers or abolishes withholding taxes on interest and dividends, or grants creditability of income taxes to thus avoiding double taxation.

Tranche. Usually a quantity or percentage of oil or gas production that is subject to specific criteria. (1) The Indonesian first tranche production (FTP) of 20% means that the first 20% of production is subject to the profit oil split and taxation, and this tranche of production is not available for cost recovery. (2) Sliding scale terms typically subject different levels of production (tranches) to different royalty rates, tax rates, or profit oil splits.

Example:

Sliding Scale Royalty:

Tranches, BOPD		Royalty, %
First Tranche	Up to 10,000	5
Second Tranche	10,001–20,000	10
Third Tranche	20,001–40,000	15
Fourth Tranche	40,001 +	20

Transfer Pricing. Integrated oil companies must establish a price at which upstream segments of the company sell crude oil production to the downstream refining and marketing segments. This is done for the purpose of accounting and tax purposes. Where intrafirm (transfer) prices are different than established market prices, governments will force companies to use a marker price or a basket price for purposes of calculating cost oil and taxes.

Transfer pricing also refers to pricing of goods in transactions between associated companies.

Treaty Shopping. Seeking out tax benefits and treaties of various countries in order to structure an appropriately situated business entity in a given country that would take advantage of benefits that would not ordinarily be available.

Turnover. A financial term that means gross revenues—used outside of the United States.

Underlifting. see Overlifting

Unit-of-Production Depreciation. Method of depreciation for capital costs. This method attempts to match the costs with the production those costs are associated with.

Formula for Unit-of-Production Method

Annual Depreciation = $(C - AD - S)\dfrac{P}{R}$

where:

C = Capital costs of equipment
AD = Accumulated depreciation
S = Salvage value
P = Barrels of oil produced during the year *
R = Recoverable reserves remaining at the beginning of the tax year

* If there is both oil and gas production associated with the capital costs being depreciated, then the gas can be converted to oil on a thermal basis.

Uplift. Common terminology for a fiscal incentive whereby the government allows the contractor to recover some additional percentage of tangible capital expenditure. For example, if a contractor spent $10 million on eligible expenditures and the government allowed a 20% uplift then the contractor would be able to recover $12 million. The uplift is similar to an investment credit. However, the term often implies that all costs are eligible where the investment credit applies to certain eligible costs. The term *uplift* is also used at times to refer to the built-in rate of return element in a rate of return contract.

Value-Added Tax. A tax that is levied at each stage of the production cycle or at the point of sale. Normally associated with consumer goods. The tax is assessed in proportion to the value added at any given stage.

Wildcat Well. An exploratory well drilled far from any proven production. Wildcat wells involve a higher degree of risk than exploratory or development wells.

Withering Clause. see Dilution Clause

Withholding Tax. A direct tax on a foreign corporation by a foreign government. The tax is levied on dividends or profits remitted to the parent company or to the home country, as well as interest paid on foreign loans.

Working Interest. The percentage interest ownership a company (or government) has in a joint venture, partnership, or consortium. The expense-bearing interests of various working-interest owners during exploration, development, and production operations may change at certain stages of a contract or license. For example, a partner with a 20% working interest in a concession may be required to pay 30% of exploration costs but only a 20% share of development costs. With government participation, the host government usually pays no exploration expenses but pays prorated development and operating costs and expenses.

World Bank. A bank (funded by approximately 130 countries) that makes loans to less developed countries (LDCs). The official name of the World Bank is the International Bank for Reconstruction and Development.

INDEX